Generative AI with SAP and Amazon Bedrock

Utilizing GenAI with SAP and AWS Business Use Cases

Miguel Figueiredo

Apress®

Generative AI with SAP and Amazon Bedrock: Utilizing GenAI with SAP and AWS Business Use Cases

Miguel Figueiredo
Sao Paulo, São Paulo, Brazil

ISBN-13 (pbk): 979-8-8688-0967-5   ISBN-13 (electronic): 979-8-8688-0968-2
https://doi.org/10.1007/979-8-8688-0968-2

Copyright © 2025 by Miguel Figueiredo

This work is subject to copyright. All rights are reserved by the Publisher, whether the whole or part of the material is concerned, specifically the rights of translation, reprinting, reuse of illustrations, recitation, broadcasting, reproduction on microfilms or in any other physical way, and transmission or information storage and retrieval, electronic adaptation, computer software, or by similar or dissimilar methodology now known or hereafter developed.

Trademarked names, logos, and images may appear in this book. Rather than use a trademark symbol with every occurrence of a trademarked name, logo, or image we use the names, logos, and images only in an editorial fashion and to the benefit of the trademark owner, with no intention of infringement of the trademark.

The use in this publication of trade names, trademarks, service marks, and similar terms, even if they are not identified as such, is not to be taken as an expression of opinion as to whether or not they are subject to proprietary rights.

While the advice and information in this book are believed to be true and accurate at the date of publication, neither the authors nor the editors nor the publisher can accept any legal responsibility for any errors or omissions that may be made. The publisher makes no warranty, express or implied, with respect to the material contained herein.

> Managing Director, Apress Media LLC: Welmoed Spahr
> Acquisitions Editor: James Robinson-Prior
> Development Editor: James Markham
> Coordinating Editor: Gryffin Winkler

Cover designed by eStudioCalamar

Cover image by geralt @ Pixabay.com

Distributed to the book trade worldwide by Apress Media, LLC, 1 New York Plaza, New York, NY 10004, U.S.A. Phone 1-800-SPRINGER, fax (201) 348-4505, e-mail orders-ny@springer-sbm.com, or visit www.springeronline.com. Apress Media, LLC is a California LLC and the sole member (owner) is Springer Science + Business Media Finance Inc (SSBM Finance Inc). SSBM Finance Inc is a **Delaware** corporation.

For information on translations, please e-mail booktranslations@springernature.com; for reprint, paperback, or audio rights, please e-mail bookpermissions@springernature.com.

Apress titles may be purchased in bulk for academic, corporate, or promotional use. eBook versions and licenses are also available for most titles. For more information, reference our Print and eBook Bulk Sales web page at http://www.apress.com/bulk-sales.

Any source code or other supplementary material referenced by the author in this book is available to readers on GitHub (https://github.com/Apress). For more detailed information, please visit https://www.apress.com/gp/services/source-code.

If disposing of this product, please recycle the paper

To my wife, son, and daughter, my greatest purpose.

Table of Contents

About the Author xi

About the Technical Reviewers xiii

Acknowledgments xv

Introduction xvii

Preface xxi

Chapter 1: Introduction to Artificial Intelligence 1
    Machine Learning 2
    Deep Learning 2
    NLP 2
    A Brief History of NLP 3
    Computer Vision 4
    The Beginning of AI 5
        Advances in Technology 9
    Current Status of AI 12
        Transforming Industries with Generative AI 13
        The Rise of Large Language Models 14
        Transformer Architecture 15
        Attention 16
        Encoders and Decoders 17
        Foundation Models 18
        Traditional ML vs. Foundation Models 18

TABLE OF CONTENTS

    Google's BERT 22

    Google's BARD 23

    Google Gemini 23

    Differences Between Google's BARD and Gemini 24

    OpenAI's GPT 25

    Midjourney 26

    OpenAI in the Spotlight 27

    The OpenAI GPT Series 29

    ChatGPT in the Spotlight 31

    Amazon AIML 32

    Amazon Bedrock 33

    Amazon Q Developer 40

    Amazon EC2 Inf2 Instances Powered by AWS Inferentia2 Chips ... 42

    New Trn1n Instances, Powered by AWS Trainium Chips 42

    SAP Joule 43

    SAP Build Code 45

    SAP AI Foundation 49

  The Future of AI 53

Chapter 2: GenAI in the Spotlight 55

  Reshaping Industries with Generative AI 55

  The Pandemic and AI Disruptions 57

    Unleashing AI as Never Before 59

    Healthcare and Biotechnology 60

    Finance and Investment 62

    Information Technology 63

    Communication Services 65

    Legal 67

    Digital Advertising 67

TABLE OF CONTENTS

Capital Goods...68
ChatGPT Spotlights GenAI 70
    A Look at the Evolution of GPT Models...71
    Ascension of Conversational AI Interfaces 74
    Pop Culture Influence 75
    Revolutionizing Linguistic Education...75
    AI Trends Shaping Pop Culture 76
    An Exciting Era Unfolds 77

Chapter 3: Opportunities and Impacts of GenAI 85
The Impact of Generative AI on Modern Industries...86
    Chest CT Images..87
    Chest X-Rays 88
    Predicting the Prognosis of COVID-19 88
    Predicting the Epidemic Trends of COVID-19..90
    Drug Discovery and Vaccine Development for COVID-19 91
    Drug Repurposing...91
    Drug Development..92
    Vaccine Development 92
Enterprise Modernizing and Faster Coding..92
    A New Hybrid World...94
    Getting Even Faster 96
    The Rise of Cyborgs...97
    The Secret Sauce 101
Prompt Engineering 102
    What Is a Prompt?...103
    What Makes Prompt Engineering So Relevant? 103
    Which Use Cases Can Take Advantage? 105
    What Are Applied Techniques? 107

TABLE OF CONTENTS

What About Best Practices? 111
Limitations and Concerns 112
    The Asking Paradigm with GenAI Models 113
    Legal Concerns 115
    Knowing Your Rights 115
    More Concerns 116

Chapter 4: Getting Started with Amazon Bedrock 127
Log in to AWS Console 127
Working with Chat Playground 133
Working with Image Playgrounds 146
Working with Image Playground (Advanced) 159

Chapter 5: Getting Started with GenAI Using SAP BTP and Amazon Bedrock 173
Log in to the AWS Console 214
Launch Amazon SageMaker Studio 217
Query SAP HANA Cloud 236
Integrate GenAI with SAP HANA Cloud 248
Experiment with Langchain SQL Agent 268

Chapter 6: Building GenAI with SAP, Lambda, and Amazon Bedrock 285
Analyze a SAP Report Using Generative AI 287
Integrate Generative AI with SAP HANA and PandasAI 300
Building Generative AI Using Lambda and API Gateway 320

Chapter 7: The AI Journey Gets Started 375
New Experiences 375
Productivity 375
Insights 376

Creativity 376
Finance Manager 378
Accounts Payable Manager 380
SAP Developer 382
Order-to-Cash and Procure-to-Pay Insights 386
AI-Powered Dashboard Authoring Experience 388
SAP Business AI 389
    ERP and Finance 391
    Human Resources 391
    Industries 392
    IT and Platforms 392
    Sourcing and Procurement 394
    Supply Chain 394
Generative AI Hub 394
SAP Interactive Value Journey 398
SAP Road Map Explorer 400
Going Beyond with Generative AI 400
    Transforming Manufacturing with AWS and SAP 401
    Working with SAP and AWS 402
    Exploring the Strategic Value of Generative AI Beyond Productivity ... 403
    Tangible Effects on Financial Operations 403
    Connecting Audiences Using a Prompt 404
    Predicting with Generative AI 404
    Redefining Customer Satisfaction 405
Last, but Not Least 406

Index **407**

About the Author

Miguel Figueiredo is a passionate software professional with more than 30 years of experience in technical solution architecture. He has a degree in information systems and an MBA from Mackenzie University, as well as an international MBA in business administration from FIA Business School in partnership with Vanderbilt University. Miguel gained his experience delivering business intelligence solutions for several Fortune 500 companies and multiple global corporations. As the SAP HANA Services Center of Excellence leader, he was responsible for the evangelization and best-practice adoption of data management and business intelligence in his region.

Currently, Miguel advises companies in maximizing value realization in their digital transformation journeys by adopting cloud initiatives. Miguel is dedicated to supporting his family and encouraging the development of good habits for health, both in body and mind.

About the Technical Reviewers

Lino Maggi is a leader with over 15 years of experience in cloud solution architecture, having worked at AWS, SAP, and Capgemini, where he sold impactful projects to Global 500 companies. With deep technical expertise, he focuses on multi-cloud digital transformations, optimizing hybrid infrastructures, and leading large-scale cloud migrations. Lino has also led teams in Generative AI projects, delivering concrete improvements in operational efficiency. His work spans industries such as Consumer Goods, Agribusiness, Manufacturing, and Mining, where he has architected ERP solutions affecting millions. Fluent in Portuguese, English, and Spanish, Lino holds an MBA in International Management and applies the discipline of a nationally-ranked karate athlete to his leadership in technology.

Felipe Pojo is a recognized leader in AI and Machine Learning at Amazon Web Services (AWS), where he oversees the AI/ML and Generative AI Go-to-Market strategy for Brazil. He holds a bachelor's degree in Business Administration, an MBA in Business Management and Information Systems, and a recent specialization in Generative AI from MIT.

ABOUT THE TECHNICAL REVIEWERS

With over nine years of experience as an AI/ML Sales Specialist, Felipe has deep expertise in areas like Conversational AI, Natural Language Processing (NLP), Voice and Speech recognition, Large Language Models (LLMs), Image Generation, and Advanced Analytics. He works with companies across sectors such as banking, telecom, and retail in Latin America, helping them develop intelligent systems that leverage data for automation and insights. His solutions enhance customer experience and drive digital transformation through technologies like contact center automation, document understanding, and predictive analytics.

Acknowledgments

A special thank you to Lino and Felipe, my partners in this incredible mission, for being responsible for a fantastic and essential technical review, enriching the book with their valuable insights. You guys are awesome!

I am eternally grateful to my awesome wife, Alessandra, for listening to the first ideas about this book and for always trusting and giving me all the support I've needed in all these years together. Thank you so much, dear.

What shall I say about my daughter Beatriz and my son Ethan? Both are extremely talented, responsible, creative, and competitive individuals who make me very proud. They encourage me to always raise the bar in my own work.

For my family: This book is dedicated to my mom Ivanir Maciel and the memory of my dad Anselmo Figueiredo for always seeking to transmit their family values, for teaching us to believe in a greater God, for inspiring us, and for dedicating a large part of their lives to us, ensuring that we could fully develop as individuals and professionals. To my sisters Selma, Monica, and Regina and my brother Max: you are always with me in my thoughts.

I want to thank *everyone* who has worked with me and contributed in any manner to my development.

A very special thank you to everyone on the Apress team who helped me so much during the entire journey. Special thanks to Divya, the ever-patient Editor; and Laura Berendson, my amazing Development Editor.

Finally, I want to express gratitude to you, the reader. Thank you! This book is for you.

Good reading!

Introduction

This book introduces generative AI and helps to develop an understanding of its key features, technology, architecture, and tangible business use cases with SAP and Amazon Bedrock. It should help you develop the skills needed to use the core features available in the SAP Business Technology Platform with Amazon Bedrock.

This book covers concepts regarding large language models (LLMs) and should equip you with practical knowledge to unleash the best use of generative AI (GenAI). As you progress, you will learn how to get started with your own LLM models, understand how to work with different models, and work with generative AI for multiple use cases.

Additionally, you will learn how to take advantage of Amazon Bedrock using AWS Lambda and API Gateway, SAP HANA Cloud, Langchain, SQL Agent, and Pandas AI.

To fully leverage this knowledge, this book will provide practical step-by-step instructions on how to establish a cloud SAP BTP account model and create your first GenAI artifacts. The book is an important prerequisite for those who want to start with generative AI with SAP and Amazon Bedrock.

This book is recommended for those who want to learn about GenAI with SAP, as it shows how to take full advantage of it and supports its practical implementation.

INTRODUCTION

This book contains the following chapters:

Chapter 1: Introduction to Artificial Intelligence
This chapter will help readers to understand new demands that have arise from the current digital transformation and how these changes affect business software needs. Along with this, it introduces how technologies like artificial intelligence (AI), natural language processing (NLP), and machine learning were disruptive during the pandemic. This chapter closes with an explanation of the status of AI.

Chapter 2: GenAI in the Spotlight
This chapter will help readers to understand what GenAI means, and discuss the popularization of the technology behind ChatGPT and OpenAI.

Chapter 3: Opportunities and Impacts of GenAI
The goal of this chapter is to help readers identify opportunities and impacts brought by generative AI in different areas.

Chapter 4: Getting Started with Amazon Bedrock
This chapter will help readers understand the playground area in Amazon Bedrock, as well as which resources and LLMs are available to start building your generative AI cases.

Chapter 5: Getting Started GenAI with SAP BTP and Amazon Bedrock
This chapter helps readers to explore the use of Amazon Sagemaker Notebook to integrate a generative artificial intelligence large language model (LLM) in order to improve the user's experience exploring their business data stored in SAP HANA Cloud.

Chapter 6: Building GenAI with SAP Report, Lambda, and Amazon Bedrock
This chapter helps readers understand how to take advantage of Amazon Bedrock to analyze SAP Report by using summarization features and leveraging Claude LLM to simulate a financial analyst. Also, it explores the capabilities of Pandas AI, SAP HANA Cloud, AWS Lambda, and API Gateway.

Chapter 7: The AI Journey Gets Started...

This chapter should help readers to envision what to expect from GenAI and offers useful resources and knowledge sources to support continuous research in the field of AI related to SAP use cases.

For the best results understanding this book, I recommend starting with Chapter 1 and reading through the book sequentially.

Preface

In July of 2022, I got started as an SAP sales specialist working at AWS with the goal of growing business related to SAP migration and modernization for many customers in different areas. It is not very well known that in January 2021, SAP launched an offering called *RISE with SAP* to help companies seize the advantages of cloud computing in their mission-critical, core systems. This offering has been important, as it accelerates the modernization of SAP customers and makes it possible to have positive discussions about modern architecture jointly using the SAP and AWS platforms.

With the first book I published, I was 100% dedicated to many agendas and situations most new employees have during the first year with a company, identifying which mechanisms would be potentially responsible for helping me achieve my goals.

The year 2022 was tremendous for me, but 2023 promised to be an even bigger deal, bringing with it new reorganization and challenging goals. Even though my focus on supporting customers in their journey stayed the same, some good opportunities presented themselves and once again I was able to achieve my goals, ending up with some good stories to share. However, within that year, the conversations around the subject of generative AI were gaining relevance that extended to the context of SAP. GTM teams were able to generate pitches and collaterals in a couple of weeks, which supported corporate messages, empowered by GenAI, seeking a change in existing business processes in pursuit of more productivity and new ways to envision and perform business functions.

PREFACE

In January 2024, many research and market analysis institutes released content about AI. A Gartner publication predicted that global spending on AI software would surge from $124 billion in 2022 to $297 billion in 2027, and the market was expected to grow at a 19.1% compound annual growth rate for the next six years.

According to this prediction, GenAI's rapid growth could be attributed to enterprise software vendors integrating AI tools into current and future releases, streamlining the widespread adoption of GenAI-based features and new apps. The same prediction holds that GenAI will eventually become a cornerstone of all AI software spending, reaching 35% of worldwide revenues by 2027.

Having reached 2024, following all the recent progress of AI, I have no doubt that all projections will continue to shape a promising scenario. This has piqued the interest of most company leaders, who wish to take advantage of GenAI capabilities with SAP and open opportunities far beyond simply migrating their SAP system to a cloud provider.

It is my belief that if you can make practical use of this essential learning to support your journey using generative AI with SAP and Amazon Bedrock, the investment you made will have paid off. Think big!

Writing a book is not an easy task, and the challenge is even harder when the subject changes so fast, as is the case with generative AI. So, don't be surprised if at the time of reading, some aspects of the topics covered have changed or been updated.

CHAPTER 1

Introduction to Artificial Intelligence

Artificial intelligence (AI) is a branch of computer science that centers on developing machines that can carry out tasks usually associated with intelligence, like voice recognition and decision-making abilities without the need for intervention or control mechanisms in place for them to operate effectively. According to John McCarthy, intelligence development is the science and engineering behind creating machines, specifically intelligent computer programs that have the capability to reason and make decisions autonomously without relying solely on observable methods, like in a science such as biology. This description encapsulates the ideas and intricacies of intelligence enough, but there a simpler explanation too. The folks at IBM put it plainly as follows: "Artificial intelligence is a blend of computer science and rich data sets used to tackle problems."

AI aims to simulate human-like thinking and decision-making processes. It involves the creation of algorithms and models that enable machines to perceive, reason, learn, and make decisions. AI can be divided into two categories: narrow AI and general AI.

- **Narrow AI**: Focuses on specific tasks and is prevalent in various applications today
- **General AI**: Refers to machines that possess human-like intelligence across a wide range of tasks

Machine Learning

Machine learning (ML) is a subset of AI that focuses on the development of algorithms and models that allow computers to learn and improve from experience without being explicitly programmed. ML algorithms analyze large datasets to identify patterns and make predictions or decisions based on those patterns. It relies on statistical techniques to automatically learn from data and adapt its performance. ML can be categorized into three types: supervised learning, unsupervised learning, and reinforcement learning.

Deep Learning

Deep learning is a subfield of ML that employs artificial neural networks to model and understand complex patterns and relationships within data. Deep learning algorithms are inspired by the structure and function of the human brain, consisting of multiple layers of interconnected nodes (neurons). These networks are capable of learning hierarchical representations of data, enabling them to extract high-level features from raw input. Deep learning has achieved significant breakthroughs in various domains, including computer vision and natural language processing (NLP).

NLP

Machines that use natural language processing (NLP) technology, a branch of AI, can interact with humans using language meaningfully by interpreting and generating human language data, for tasks such as text classification and sentiment analysis, language translation, question answering, and speech recognition. NLP combines methods, from the fields of linguistics, machine learning, and deep learning to handle and examine text data effectively.

A Brief History of NLP

The genius of Alan Turing's work sparked German-American computer scientist Joseph Weizenbaum's interest in developing a computer program capable of reading our interactions, understanding what we are saying, and replicating responses conforming to human expectations.

Between 1964 and 1966, working at the Artificial Intelligence Laboratory at the Massachusetts Institute of Technology (MIT), Weizenbaum developed the ELIZA computer program, whose objective was to make users believe they were talking to a human being. ELIZA was designed to mimic a therapist who would ask open-ended questions and respond in natural language.

The designation *ELIZA* was a tribute to the character from George Bernard Shaw's *Pygmalion* named Eliza Doolittle, a hard-working heroine who learns to speak in an upper-class accent.

The program had ready-made responses to make users believe that behind the machine, a therapist was analyzing the situation and carrying on a formal conversation. These responses were triggered by keywords that the program had previously received as input, returning phrases related to these keywords to the user who used them.

Because of the rudimentary form of computers at the time, the user interacted remotely through an electric typewriter, writing a sentence that was sent to the mainframe computer where ELIZA was installed. The message was examined; ELIZA looked for the keywords and returned the answer in sentences. The user then received a printed response. The idea of turning ELIZA into a therapist was to encourage a conversation, an ongoing dialogue between the user and the machine. In fact, Weisenbaum called the set of scripts programmed for ELIZA "DOCTOR."

CHAPTER 1 INTRODUCTION TO ARTIFICIAL INTELLIGENCE

For example, if the user typed the phrase, "I have problems with my parents," ELIZA might respond, "Tell me about your family."

Figure 1-1 shows a fictional conversation with the ELIZA program.

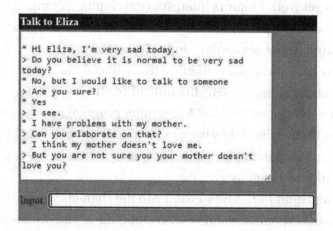

Figure 1-1. *Talking to Eliza about family problems*

Computer Vision

Computer vision is about creating algorithms and models that help machines interpret information from images or videos like humans by extracting important insights from the data and using it to make decisions or take actions in various tasks such as identifying objects in images and videos or analyzing visuals through methods, like image processing and deep learning.

In summary, AI is the broader concept of creating intelligent machines, while ML, deep learning, NLP, and computer vision are specific subfields within AI.

CHAPTER 1   INTRODUCTION TO ARTIFICIAL INTELLIGENCE

The Beginning of AI

The beginning of AI came in the 1950s when the first computers and programs capable of simulation, logical reasoning, and learning appeared. It was at this time that Alan Turing proposed the famous Turing test to see if a machine could pass as a human in a conversation (see Figure 1-2).

In Turing's article "Computing Machinery and Intelligence," he explains, "These questions replace our original, 'Can machines think?'"

In his paper discussing the matter of whether machines can engage in thought processes or not Turing suggests that due to the definitions of the terms "think" and "machine," it would be more beneficial to rephrase the question with a different one that is closely connected but can be expressed more clearly. First, he needs to identify a clear concept to substitute the term "think." The next step is to specify the types of "machines" he has in mind, and lastly, armed with these strategies, he crafts a query linked to the initial one that he feels confident he could respond positively to.

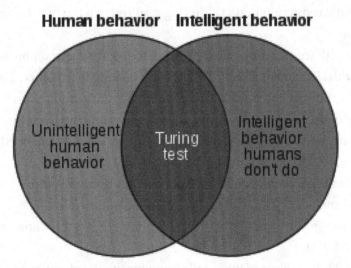

Figure 1-2. *Venn diagram on Turing test*

5

CHAPTER 1 INTRODUCTION TO ARTIFICIAL INTELLIGENCE

In the same period, McCarthy coined the term *artificial intelligence* at the Dartmouth Conference (1956). According to McCarthy, "Each aspect of learning or any other form of intelligence can be described so precisely that a machine can be created to simulate it."

McCarthy was also the creator of the Lisp programming language widely used in AI and the concept of time-sharing, which allowed the simultaneous use of the same computer by several users.

Other relevant names in the beginning of AI were Marcin Minsky, considered the father of modern AI, who founded the MIT Artificial Intelligence Laboratory in 1959; Claude Shannon, who applied information theory to cryptography and chess; and Herbert Simon and Allen Newell, who developed the first program capable of solving logic problems, the science of communication and control between men and machines.

This period of AI was great in optimism and creativity but challenging due to its limitations. AI pioneers dreamed of intelligent machines able to outperform humans in many areas but recognized the obstacles and risks. The dramatic reduction in R&D funding resulted in a fragmented scientific community.

Along with this scenario, the hardware and software available were limited, with slow and expensive computers lacking in memory. The simple algorithms could not deal with ambiguity, uncertainty, and variability of actual data. Later, other challenges came up because scaling, maintaining, and adapting AI programs to new domains were needed.

Almost two decades later, new technological and theoretical advancements allowed R&D in artificial intelligence to resume. The emergence of artificial neural networks, machine learning, distributed computing, and the Internet opened new possibilities and challenges for AI.

The 1980s brought a historic milestone for the development of AI to be seen as not just an academic research area but as a practical reality, capable of solving complex problems and challenging human intelligence.

CHAPTER 1 INTRODUCTION TO ARTIFICIAL INTELLIGENCE

Factors like the progress of computer science, the increase in computer processing and storage capacity, and the rise of new sources of funding and support from governments and companies brought the AI revolution into the spotlight.

In the AI revolution's evolution, key players included researchers and developers who worked to expand its horizons using methods and models, beyond the barriers of logic and rigid rules that had limitations in handling complex and unstructured domains effectively. These new techniques and paradigms emerged to overcome those limitations and project AI to handle a more complex domain where uncertainty, ambiguity, and a dynamic situation were present.

> **Artificial neural networks (ANNs)**: Computational models inspired by the functioning of the human brain formed by interconnected processing units that learn from data and experiences. ANNs proved to be efficient in performing tasks such as pattern recognition, classification, prediction, and optimization.
>
> **Evolutionary computation (EC)**: An area that uses the principles of evolutionary biology, such as natural selection, mutation, and recombination, to generate adaptive solutions to complex problems. EC encompasses techniques such as genetic algorithms, evolutionary strategies, genetic and evolutionary programming, and ant colonies.
>
> **Expert systems (ESs)**: Programs that simulate the reasoning of a human expert in a specific domain through a knowledge base and an inference engine. ESs were widely used to support decision-making processes in areas such as medicine, engineering, law, and education.

CHAPTER 1   INTRODUCTION TO ARTIFICIAL INTELLIGENCE

Intelligent agents (IAs): Autonomous entities that perceive the environment, reason through their goals, and act to achieve them. IAs can be simple or complex, individual or collective, and cooperative or competitive, and they can act in virtual or real environments such as electronic games or robotics.

The emergence of these new techniques and paradigms allowed AI to be applied to broader and more sophisticated domains, such as voice and image recognition, automatic translation, electronic games, robotics, and medicine. The AI revolution gained even more momentum in the 2000s with the advent of the Internet, big data, and deep learning, which took AI to a new level of performance and popularity.

The Internet has opened up amounts of data that serve as the foundation for AI algorithms to operate on effectively. Big data encompasses the methods and technologies that facilitate the storage and analysis of data to derive insights for purposes. Deep learning stands out as a sector within AI that leverages networks modeled after the human brain to grasp information from data and accomplish intricate duties, like identifying images and understanding speech and language.

These advanced technologies have enabled the integration of AI across sectors, such as healthcare, education, security, entertainment, and transportation. AI has been instrumental in enhancing the quality of life boosting productivity and facilitating communication. However, the rise of AI also poses challenges and risks including social economic and legal dilemmas. For instance, how do we ensure that accountability is maintained to uphold human rights privacy and security when using AI? What are some ways to address the effects of AI on jobs and education or its impact on culture? How can we make sure that AI remains transparent and fair while maintaining reliability?

These are the sort of challenges that society, government, companies, and researchers involved with AI must discuss. The AI revolution is a reality that cannot be ignored or feared. We should instead attempt to understand it and take advantage of it consciously and ethically.

Advances In Technology

Despite the sudden popularity of ChatGPT and generative AI technologies we've seen recently, their gradual evolution has indeed been based on layers of extensive research and constant development in the field over the past *seven decades*. The following are some of the fundamental discoveries and advances that led to the creation of increasingly sophisticated AI systems:

- The invention of the first digital computer and programming language, which allowed the implementation of algorithms and symbolic methods to solve logical and mathematical problems

- The emergence of cybernetics and information theory, which provided the concepts of feedback, communication, and control to model complex and adaptive systems

- The development of fuzzy logic and expert systems, which enabled the representation and treatment of imprecise, uncertain, and incomplete knowledge

- The creation of artificial neural networks and genetic algorithms, which were inspired by biology to create systems capable of learning and evolving from data and experiences

CHAPTER 1   INTRODUCTION TO ARTIFICIAL INTELLIGENCE

- The introduction of machine learning and deep learning, which use statistical and computational techniques to extract patterns and knowledge from large volumes of data

- The emergence of artificial general intelligence and superhuman artificial intelligence, which aim to create systems with cognitive capacities equivalent to or superior to those of humans in all areas of knowledge

- The application of artificial intelligence in different domains and sectors, such as health, education, manufacturing, commerce, entertainment, and security, which has generated both benefits and challenges for society

Among the most recent and promising AI technologies is generative AI, a type of system capable of generating original and realistic content from data or instructions, such as text, images, videos, and audio. There are examples of this technology that can generate coherent and creative dialogues on various topics using new and diverse language models based on deep neural networks.

Another area that has used AI widely in digital transformation is data analytics. With the increasing amount of data that companies generate, the analytical process has become complex and time-consuming. AI can help automate data analysis, helping to identify patterns and trends to provide valuable insights for decision-makers. Marketing departments can use data to create and deliver personalized content, customize experiences to each person according to their behavior on the network, make recommendations, and facilitate the connection between companies and their customers.

AI can also be used to create virtual assistants, chatbots, and other customer service solutions. With AI, companies can offer personalized and efficient support to their customers, improving customer satisfaction and reducing customer service costs. The effective use of AI to build immersive creative experiences, involving anything from augmented reality (AR) to virtual reality (VR), leads to increased engagement online and in physical social networks. As a result, customer behaviors can be popularized to increase sales conversion. The opposite is also true. The misuse of these technologies and mainly poor planning in their use can discredit brands and cause them to lose market share.

It's crucial to note that incorporating AI into interactions comes with its set of obstacles and worries as well. The protection of user data privacy is a topic that's garnering attention and discussion in recent times. Businesses have the responsibility to gather and utilize data in an ethical ly sound manner while also being transparent and ensuring security measures. If data were to fall into the wrong hands, it could lead to severe consequences for companies. Nevertheless, leveraging this technology appropriately has the potential to yield benefits in return.

For AI to be seen as a tool for the common good, which can contribute to human progress and solve global problems, researchers, developers, users, regulators, and other actors involved in AI must follow ethical and legal principles that promote trust, responsibility, and inclusion.

In the seven decades or so artificial intelligence has made strides bringing about digital changes and crafting unique digital interactions tailored to customers in different sectors. AI is set to keep changing the world molding the future in fields like healthcare, finance, and transportation.

We must use it responsibly and ethically to advance humanities's progress by emphasizing the importance of AI technologies in AI's evolution process as they enable systems to learn and innovate from available data sources to produce more advanced and unexpected outcomes.

As generative AI evolves, new possibilities for creating content, solving complex problems and even creating new forms of art are emerging. Through the use of advanced technologies such as generative AI, as long as people's data is shared with their consent, AI will become mature enough to transform the world we live in and shape the future of all industries. The possibilities are countless, and those who will know how to apply AI into right opportunities will reap maximum benefits.

Current Status of AI

Artificial intelligence is a changing field that has seen progress over time and introduced new tools and applications that are reshaping different industries and sectors today. Lately there has been an increasing emphasis placed on three areas within AI: general intelligence (AGI) generative AI (GenAI), and large language models (LLMs). These areas represent cutting-edge research in AI and are fueling innovation and transformation in their relevant sectors.

AGI is a subdomain of AI that aims to create systems that can perform any intellectual task that humans can—with the potential to even surpass human intelligence. Whereas current AI systems are specialized and designed to perform specific tasks, AGI systems would be capable of learning and adapting like humans to new tasks and situations. AGI has been the subject of much research and debate in the AI community, and while we still may be far from achieving true AGI, progress in this area could have profound implications for humanity.

There are currently no existing AGI systems, as their creation remains in the realm of research and development. However, many examples of specialized AI systems have achieved human-level performance in specific tasks. For instance, OpenAI's GPT-3 model launch demonstrated how foundation models (FMs) are able to generate text indistinguishable from that written by humans, as did many other FMs that came to life after ChatGPT, such as Anthropic, Mistral, and Meta.

Even though AGI is still a theoretical concept, there are already powerful platforms and tools pushing the boundaries of what is possible. Using generative AI and LLMs, they've been focusing on the development of the underlying technologies required for AGI. A notable platform in the AGI space is DeepMind, a research lab acquired by Google in 2015. DeepMind has made significant strides in developing AI systems capable of learning and adapting to new situations similarly to humans. For example, the AlphaGo system designed by DeepMind has defeated multiple human world champions in the game of Go, a feat previously thought impossible for AI systems.

While artificial general intelligence aims to create machines that can learn and adapt to new situations at a human level, generative AI focuses on creating new content with minimal human input. Despite their distinct goals, both subdomains of AI are driven by the broader goal of creating machines that can demonstrate more autonomy and creativity in their decision-making processes. In fact, some experts believe that the development of generative AI is a crucial step toward achieving AGI, as machines that can generate their own content and ideas are more likely to learn and adapt to match human expectations.

Transforming Industries with Generative AI

Generative AI is one of the most promising subdomains of artificial intelligence; it has gained significant traction in recent years and holds immense potential for transforming a wide range of industries and domains. It involves using machine learning algorithms to create new content, such as images, music, and text that are similar in style and quality to the input data. By analyzing large sets of existing data, these algorithms can generate new and unique content, opening exciting possibilities for new forms of creativity and expression. For instance, an AI system trained on images of faces could generate photorealistic images of faces that have never existed before.

Various applications have already used generative AI for anything from visual art and music composition to literature and video production. A growing number of platforms and tools are available that make it more accessible to developers, artists, and creatives.

CHAPTER 1   INTRODUCTION TO ARTIFICIAL INTELLIGENCE

The Rise of Large Language Models

AI has advanced considerably in the realm of large language models. LLMs are artificial intelligence systems capable of comprehending and manipulating human language, which is a branch of natural language processing that delves into teaching computers how to grasp and produce human language effectively by employing methods, like machine learning and deep neural networks, for language analysis and generation.

LLMs undergo training using datasets sourced from materials, like books and online content, to grasp the nuances and organization of human language effectively in a process known as *language modeling*. The primary objective of language modeling is to educate LLMs in foreseeing the probability of word sequences by leveraging the patterns extracted from their training data. After the LLM is trained successfully, it can be applied for purposes, like translating text categorizing text content summarizing information generating text and analyzing sentiments.

The Transformer architecture introduced in a 2017 paper by Vaswani et al. depicted in Figure 1-3 stands out as an advancement in the development of LLMs. This model brought about a paradigm shift in natural language processing by employing a mechanism that enables it to focus on segments of the input sequence while encoding information.

This attention mechanism enables the model to learn the relationships between different words in a sequence or between different sequences in a document. This significantly improves the model's ability to capture long-range dependencies in the input sequence. This has led to significant improvements in performance on a wide range of language tasks, making the model much more efficient at understanding and generating natural language. Transformers have greatly expanded the capabilities of LLMs and opened new possibilities for natural language processing tasks. The Transformer architecture has since become foundational for many state-of-the-art LLMs, including OpenAI's GPT (generative pretrained transformer) series and Google's BERT (bidirectional encoder representations from transformers).

CHAPTER 1   INTRODUCTION TO ARTIFICIAL INTELLIGENCE

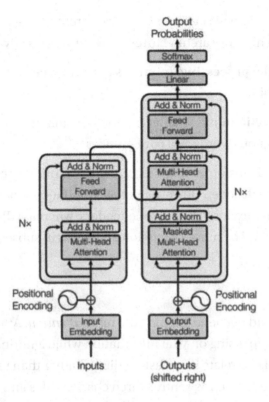

Figure 1-3. *The Transformer architecture*

Transformer Architecture

Most foundation models employ the Transformer architecture.

In 2017, transformers were introduced. A *transformer* is a deep learning model that uses the self-attention mechanism to weigh the relevance of each input data point differently. Its primary applications are natural language processing and computer vision.

Transformers, the next generation of recurrent neural networks and long short-term memory architectures, provide various advantages:

- **Parallel processing**: Improves performance and scalability
- **Bidirectionality**: Enables comprehension of ambiguous words and coreferences

The original Transformer architecture has two primary components: an encoder and a decoder. However, not all foundation models have both pieces. BERT exclusively utilizes encoders, whereas GPT uses only decoders. There will be more discussion about this in subsequent chapters.

Attention

Both encoders and decoders employ the term *attention*. Attention, in essence, implies focusing on vital information while ignoring irrelevant information. I like to relate it to "fast reading." Rather than reading entire articles or even books, I frequently search chapter titles and the first words of paragraphs and go through paragraphs for keywords to identify what I'm looking for.

The words in an article, the sections of a picture, or the words in a phrase that should receive the most attention vary according to what you are looking for.

Let's look at a simple example statement: "Beatriz went to a traditional restaurant to meet her boyfriend that night."

The following terms should draw your attention to the following queries:

- **What?** *went, meet*
- **Where?** *a traditional restaurant*
- **Who?** *Beatriz, her boyfriend*
- **When?** *that night*

Encoders and decoders in transformers employ *queries*, *keys*, and *values* to determine the attention of words (or tokens). All of them are shown as vectors. Certain queries provide keys that are closest to the query vector. Keys are encoded representations of values; in basic circumstances, they might be the same.

There are several algorithms for implementing the attention notion. I believe a simple method to understand how this works is to rank terms that are frequently used together in phrases. For example, *where* and *restaurant* are probably more closely related than *restaurant* and *faith*. So, for the query *where*, the word *restaurant* receives greater attention.

Encoders and Decoders

As previously stated, there are encoders and decoders. BERT solely employs encoders, whereas GPT uses only decoders. Both approaches recognize language, including syntax and semantics. The next generation of huge language models, such as GPT, which has billions of parameters, performs particularly well.

The two models address distinct circumstances. However, because the subject of *foundation* models is evolving, the distinctions are often fuzzy:

- **BERT (encoder)**: Classification (*e.g.*, sentiment), questions and answers, summarization, named entity recognition

- **GPT (decoder)**: Translation, generation (*e.g.*, stories)

The outputs of the *core* models are different:

- **BERT (encoder)**: Embeddings that represent words with attention to information in a certain context

- **GPT (decoder)**: Next words with probabilities

Both models are pretrained and may be utilized without requiring further training. Some of them are open source and may be downloaded from sites like Hugging Face, while others are commercial. Reuse of models is vital, since training is generally resource-intensive and costly, which few businesses can afford.

Pretrained models may be modified and adjusted for various domains and purposes. Layers can occasionally be reused without any changes, and further layers can be put on top. If layers must be updated, the new training becomes more costly. The approach for customizing these models is known as *transfer learning*, because the same general model may simply be applied to various domains.

Foundation Models

BERT and GPT are both foundation models. Let's look at their definition and attributes:

- They are pretrained on a variety of unlabeled datasets, including language and pictures.

- The self-supervised learning uses generalized data representations for downstream tasks like classification and generation.

- The Transformer architecture is commonly used, but not mandatory.

Traditional ML vs. Foundation Models

Foundation models have been largely used with pretrained data provided, which allows for faster and wider applications (Figure 1-4).

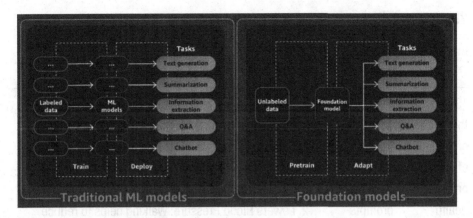

Figure 1-4. *Bedrock using Anthropic's Claude model*

A few types of foundation models have been introduced leveraging pretrained data into FMs, as you can see in Table 1-1.

CHAPTER 1   INTRODUCTION TO ARTIFICIAL INTELLIGENCE

Table 1-1. *Foundation Model Examples*

Input	Foundation Model	Output
"Summarize the articles on impact of walking on heart health"	Text-to-text Generate text from simple natural-language prompts for various applications	"Walking has a significant positive impact on heart health, offering numerous cardiovascular benefits: 1. Improves Cardiovascular Fitness: Regular walking strengthens the heart, improving its efficiency and ability to pump blood throughout the body. 2. Lowers Blood Pressure: Walking helps to reduce hypertension by enhancing circulation and reducing arterial stiffness. 3. Reduces Bad Cholesterol: It decreases LDL (bad cholesterol) levels while potentially increasing HDL (good cholesterol). 4. Improves Circulation: Walking enhances blood flow, reducing the risk of clots and improving overall vascular health. 5. Weight Management: By aiding in weight control, walking reduces strain on the heart and decreases the risk of heart disease. 6. Lowers Risk of Heart Disease and Stroke: Studies show that consistent walking can lower the risk of coronary artery disease and stroke by improving overall cardiovascular health. 7. Stress Reduction: Walking can reduce stress, which is a significant factor in heart health, by promoting relaxation and lowering stress-related hormone levels. A moderate pace of 30 minutes a day, at least five times a week, is sufficient for these benefits. Consistency is key for long-term heart health."

(*continued*)

CHAPTER 1　INTRODUCTION TO ARTIFICIAL INTELLIGENCE

Table 1-1. (*continued*)

Input	Foundation Model	Output
"Hand soap"	Text-to-embeddings Generate numerical representation of text for applications like search and finding similarities between documents	Pre-trained embeddings like Word2Vec, GloVe, or FastText map words into dense vector spaces. Example (using Word2Vec): from gensim.models import KeyedVectors model = KeyedVectors.load_word2vec_format('GoogleNews-vectors-negative300.bin', binary=True) hand_vector = model['hand'] soap_vector = model['soap'] combined_vector = (hand_vector + soap_vector) / 2 # Averaging vectors This would yield a dense vector, e.g., [0.134, -0.253, 0.678, ..., -0.124] (300 dimensions).
"An image of an astronaut riding a horse on Jupiter"	Multimodal Generate and edit images from natural-language prompts	

The FMs are being used largely in different industry applications, as follows:

- **Communications**: Chatbot, question answering, search
- **Healthcare**: Protein folding, drug development, personalized medicine, improved medical imaging

21

- **Media and entertainment**: Video game generation, upscaling content, face synthesis, film preservation and coloring

- **Automotive**: Autonomous vehicles, design parts for fuel efficiency

- **Financial services**: Risk management, fraud detection, document analysis, financial analysis

- **Consumer goods**: Optimize pricing and inventory, correct flagging of product brand and category

- **Energy & utilities**: Design renewable energy sources for generative engine optimization (GEO), predictive maintenance

- **Technology hardware**: Chip design, robotics

Google's BERT

BERT is a pretrained LLM that has gained significant attention in recent years. Developed by Google in 2018, BERT uses a bidirectional approach to language modeling, allowing it to understand the meaning of words in a sentence. This means BERT can get the meaning of a word based on the words that come before and after it in a sentence rather than just relying on the immediate context. BERT is trained on massive amounts of text data, such as Wikipedia articles and books, and is able to generate high-quality representations of language that can be fine-tuned for specific tasks. BERT has been used for a wide range of applications, such as improving search results, natural language understanding in chatbots, and even language translation. Its success has led to the development of other large-scale pretrained LLMs, such as GPT-2 and RoBERTa.

Google's BARD

Google BARD is a language model that has been extensively trained on a range of textual and coding data sources. It is capable of creating text content across genres and languages and providing responses to queries. Known for its effectiveness in research applications such as data analysis and experiment planning, this formidable tool has proven invaluable in expediting the progress of investigations.

At its core, Google BARD uses search algorithms and natural language processing techniques to help researchers find the specific data and resources they need. Researchers can search for data by entering keywords or phrases related to their research interests, and BARD will return a list of relevant datasets, publications, and other resources.

Using BARD can enhance productivity and creativity while piquing curiosity well! For example, BARD could offer suggestions to help someone achieve their reading goals for the year or break down the world of quantum physics in an easy-to-understand way.

Google Gemini

Gemini is a Google-developed AI system that came after BARD. Gemini specializes in creating varied images using generative model techniques, like generative adversarial networks (GANs). Its main objective is to advance image synthesis capabilities for use in artistry and entertainment by exploring horizons and possibilities through approaches.

CHAPTER 1   INTRODUCTION TO ARTIFICIAL INTELLIGENCE

Differences Between Google's BARD and Gemini

These are the differences between the two:

Focus: BARD automates the machine learning pipeline, which includes architecture search and hyperparameter tweaking. In contrast, Gemini is focused on image synthesis and the creation of visually pleasing and diversified pictures.

Techniques: BARD uses reinforcement learning to optimize model architectures and hyperparameters, whereas Gemini uses GANs for high-fidelity picture generation.

Application: BARD's automated machine learning skills identify applications in a variety of business domains that need successful model creation. Gemini, on the other hand, focuses on the creative domain, giving artists, designers, and content makers tools for creating distinctive and realistic visual material.

Output: BARD produces trained machine learning models that are ready for deployment, whereas Gemini provides pictures and visual material suited to individual requirements.

Scalability: Although BARD and Gemini are meant to scale, their scalability requirements differ. BARD works with complex datasets to find the best models, whereas Gemini generates diverse and high-fidelity pictures.

In 2024, Google unified BARD and Duet AI under the Gemini brand.

OpenAI's GPT

OpenAI's GPT series is another example of the LLMs that have gained significant attention and use in recent years. The GPT series comprises language models that employ deep learning methods to examine and produce human language expressions effectively. They undergo pretraining using text datasets, like books and online content from websites and social media platforms. They can then be customized for purposes such as translating languages, summarizing texts, and answering queries.

LLMs have found applications beyond machine translation systems. They are now being utilized in various other areas that involve understanding and processing human language effectively. One notable example is the use of LLMs in chatbots and virtual assistants where they help interpret user inquiries and formulate responses. These systems employ natural language processing methods to grasp the meaning and context of user questions before utilizing LLMs to generate precise replies. Chatbots and virtual assistants are seeing increased adoption in fields like customer service and healthcare, for delivering effective assistance to users. Nevertheless, the applications built on LLM technology encounter obstacles, like bias and privacy concerns, that need to be tackled. Utilizing LLMs in these contexts marks a progression in creating AI systems capable of comprehending and engaging with humans using everyday language.

Several powerful tools and platforms are available for developers and researchers who want to work with LLMs.

The fusion of AI technology with AGIs along with the utilization of LLMs holds the promise of ushering in impactful and groundbreaking applications across diverse sectors and industries. One potential application could involve AGIs leveraging datasets to generative AI capabilities for developing solutions and driving innovations within realms like healthcare research and development as well, as scientific and engineering advancements.

The proposed strategies could be converted into languages through LLMs enabling their global dissemination and application worldwide. Additionally, the utilization of LLMs can enhance the capabilities of AI systems by analyzing and producing natural language explanations for the content generated by systems thereby offering users more comprehensive information and context.

OpenAI's GPT series is an excellent example of how these different subdomains of AI can work together to create powerful and innovative applications. The GPT models are LLMs that are pretrained on massive amounts of data and generate human-like text. The models use the Transformer architecture, a neural network architecture specifically designed for natural language processing tasks, such as language translation and modeling. The GPT series represents a significant advancement in generative AI and LLMs, as they can generate high-quality text that is almost indistinguishable from human writing. Additionally, GPT-3 has been praised for its ability to perform language-related tasks previously thought to be exclusive to humans, such as writing essays, generating poetry, and even writing code. This shows how the combination of generative AI, LLMs, and AGI can lead to breakthroughs in AI and create applications once thought impossible.

Midjourney

Midjourney is an intelligence program and service developed by the independent research lab Midjourney Inc., headquartered in San Francisco. The program creates images based on descriptions in language known as prompts, a concept used in DALL E by OpenAI and Stable Diffusion by Stability AI.

Midjourney uses two very new machine learning technologies: big language models and diffusion models. You may be familiar with the former if you've utilized generative AI chatbots like ChatGPT. Midjourney initially uses a huge language model to grasp the meaning of the words you put into its prompts. This is then transformed into a vector, which is essentially a numerical representation of your prompt. Finally, this vector contributes to another complicated process called diffusion. Midjourney uses a diffusion technique to transform random noise into beautiful artwork.

Diffusion has just gained popularity in the last decade or two, explaining the rapid influx of AI-generated art. A diffusion model involves a computer progressively adding random noise to its training dataset of pictures. Over time, it learns to retrieve the original image by reversing the noise. The theory is that with enough training, such a model may learn to create totally new pictures.

OpenAI in the Spotlight

One of the most influential organizations in this space is OpenAI, a research lab founded in 2015 to create AGI that benefits humanity. OpenAI has been at the forefront of many advances in generative AI, developing models and platforms that have pushed the boundaries of what is possible.

A notable tool from OpenAI is DALL-E, a generative AI model capable of creating images simply by describing what they want in words. For example, users could request a "monkey with sunglasses," and DALL-E would generate an image that fits that description (see Figure 1-5). This technology can potentially revolutionize fields such as graphic design, advertising, and entertainment.

CHAPTER 1   INTRODUCTION TO ARTIFICIAL INTELLIGENCE

Figure 1-5. *DALL-E-generated image*

In addition, OpenAI has created several platforms and resources that make it easier for developers and researchers to work with generative AI. One such platform is OpenAI Gym, a toolkit for developing and comparing reinforcement learning algorithms. Another is OpenAI Codex, a generative AI model capable of generating code based on natural language input. OpenAI has also released several research papers and open-source projects, making it easier for others in the field to build upon their work.

OpenAI has certainly made an impact in the realm of AI with its remarkable creation GPT-3. It uses an advanced language model that excels in producing logical and contextually appropriate text based on specific prompts. It has been employed to craft forms of content like articles, research papers, and poetic works, lauded as a progression in the domain of natural language processing. OpenAI has made GPT-3 available through different platforms and tools, including Hugging Face Transformers and EleutherAI's GPT-Neo, making it more accessible to developers and researchers worldwide.

The OpenAI GPT Series

GPT is a series of language models developed by OpenAI that have been at the forefront of many advances in natural language processing. The models are based on the Transformer architecture, a neural network architecture introduced in a 2017 research paper. This architecture is particularly well-suited for language processing tasks, as it can handle long text sequences.

The first version of GPT, GPT-1, was released in 2018 and contained 117 million parameters. The model was trained on a large corpus of text data, allowing it to generate coherent and contextually relevant text in response to a given prompt. GPT-1 was a significant step forward in natural language processing and helped to establish the Transformer architecture as a powerful tool for language modeling.

In 2019, OpenAI released GPT-2, a much larger and more powerful model with 1.5 billion parameters. GPT-2 could generate even more coherent and contextually relevant text than its predecessor, and its release caused a significant amount of controversy due to concerns over its potential misuse. OpenAI initially limited access to the model, citing concerns over its potential to generate fake news and advertising. However, the model was eventually released to the public in a modified form.

ChatGPT, version 3 of GPT (GPT-3), was released in 2020 and has become immensely popular. It contains a staggering 175 billion parameters. GPT-3 can generate text almost indistinguishable from text written by humans and has been used to create content such as news, articles, essays, and even poetry. The model has been hailed as a significant step forward in natural language processing and has generated a great deal of excitement and interest within the AI community.

In March 2023, OpenAI released GPT-3.5-Turbo. The GPT-3.5-Turbo model could accept a series of messages as input, unlike the previous version, which allowed only a single-text prompt. This capability unlocked some exciting features, such as the ability to store as context previous responses and queries with a predefined set of instructions. It was the most capable and cost-effective model, which has been optimized for chat but works well for traditional completion of tasks as well.

In April 2023, OpenAI started the gradual rollout of GPT4, which was made available through a waitlist, though ChatGPT users could already access it through the Plus Account. GPT-4 is a large multimodal model (accepting image and text inputs and emitting text output) that, although less capable than humans in many real-world scenarios, exhibits human-level performance on various professional and academic benchmarks. The difference between GPT-4 and GPT-3.5 is exhibited when the complexity of a task reaches a sufficient threshold. GPT-4 is more reliable, creative, and capable of handling much more nuanced instructions than GPT-3.5.[1]

One important aspect to note about the OpenAI GPT series is that it has not only advanced the field of LLMs but has also opened new opportunities for conversational AI. ChatGPT, for example, is a large-scale generative model based on the GPT architecture designed explicitly for chatbot applications. It is trained on massive amounts of conversational data and can generate contextually relevant and natural-sounding responses. By combining the advancements made in LLMs and generative AI with a focus on natural language conversation, ChatGPT represents a promising step toward creating more intelligent and human-like chatbots. As these technologies continue to develop, we can expect to see chatbots capable of carrying out more complex tasks and engaging in more meaningful interactions with humans.

---

[1] In 2024, OpenAI released GPT-4o, which followed the trend of multimodal models not only able to process and generate text but also read images, understand audio, and generate text, images, and audio.

CHAPTER 1   INTRODUCTION TO ARTIFICIAL INTELLIGENCE

ChatGPT in the Spotlight

ChatGPT is a specific implementation of the GPT architecture designed for use in conversational AI applications, such as chatbots. ChatGPT is based on the GPT-3.5-Turbo model and has been fine-tuned on a large dataset of conversational text, which includes dialogues from social media, customer service interactions, and other sources. This fine-tuning process enables ChatGPT to generate contextually relevant and appropriate responses to any given conversation. Released on November 30, 2022, as a proof of concept, it gained an impressive 1 million users in five days and blasted into the collective consciousness.

ChatGPT is capable of answering types of inputs like questions and statements and can even generate subsequent replies to previous messages to enable more in-depth interactions in conversations. The model can also be tailored to offer responses based on fields like finance, healthcare, or education.

One of the main advantages of ChatGPT is that it can be easily integrated into existing chatbot platforms and messaging applications, such as Facebook Messenger, WhatsApp, and Slack, making it easier for businesses and organizations to add conversational AI capabilities to their existing customer service or support channels.

Despite its many advantages, ChatGPT has limitations. Like all language models, it can produce biased or inappropriate responses if trained on biased and inappropriate data. Additionally, ChatGPT might struggle to understand nuances of human language, such as sarcasm or irony, and generate responses considered inappropriate or off-topic. Finally, ChatGPT is not yet capable of passing the Turing test, which determines whether an AI system is indistinguishable from a human in conversation.

Overall, ChatGPT represents a significant step forward in the development of conversational AI, and it has the potential to revolutionize how we interact with machines. As technology continues to develop and mature, we can expect to see even more innovative chatbot platforms and applications emerge, making it even easier for businesses and organizations to incorporate conversational AI into their operations.

Amazon AIML

Amazon believes that AI and ML are among the most transformative technologies of our time, capable of addressing some of humanity's most challenging issues. That is why, over the last two decades, the company has made significant investments in the development of AI and machine learning, incorporating these capabilities into every business unit.

Today, Amazon Web Services ML solutions are working on behalf of hundreds of millions of Amazon customers worldwide, adding real value by reducing friction in supply chains, customizing digital experiences, and making goods and services more accessible and affordable. More than 100,000 companies run AI and ML workloads in AWS's cloud.

For a long time, AI has powered many everyday experiences. When you ask Alexa to play a song, walk out of an Amazon Just Walk Out store with a sandwich in your hand, or click Play on an Amazon Prime movie recommendation, AI is present. More specifically, these experiences often involve interactions with ML models.

Amazon has used generative AI in other ways:

> **Amazon Search**: Utilizes pretrained FMs to improve search results on Amazon.com
>
> **Alexa Teacher Model**: Exceeds other large language models on few-shot tasks such as summarization and machine translation

Amazon Q Developer [2]: Helps developers build applications faster and more securely with an AI coding companion

AWS has collaborated with Hugging Face to easily fine-tune and deploy next-generation ML models on EC2 and Amazon SageMaker. It has also been an integral partner in scaling Stability AI open-source foundation models across modalities.

Amazon Bedrock

Amazon Bedrock became widely accessible on September 28, 2023. It serves as a managed service designed for creating and expanding generative AI applications that can produce text, images, audio, and synthetic data upon receiving prompts.

Along with many features, Amazon Bedrock brings key benefits allowing customers to do the following:

- Accelerate development of generative AI applications using FMs through an API without having to manage infrastructure

- Choose FMs from AI21 Labs, Anthropic, Cohere, Meta, Mistral, Stability AI, and Amazon to find the right FM for a use case

- Privately customize FMs using an organization's data

- Enhance data protection using comprehensive AWS security capabilities

- Use familiar AWS tools and capabilities to deploy scalable, reliable, and secure generative AI applications

---

[2] As of April 30, 2024, Amazon CodeWhisperer is now part of Amazon Q Developer including inline code suggestions and Amazon Q Developer security scans.

CHAPTER 1   INTRODUCTION TO ARTIFICIAL INTELLIGENCE

The customer can begin with an FM from Amazon or a third-party AI business via an API and then build up from the pretrained "bedrock" of the basic model to adapt it to their specific needs.

Because it is completely managed, the customer doesn't need to handle the infrastructure, and the FM can be privately personalized with an organization's data. There is a broad range of supported FM providers to choose from.

Amazon Bedrock provides easy access to FMs, which are ultra-large ML models used in generative AI from top AI start-up model providers such as AI21 Labs, Anthropic, Cohere, Meta, Mistral, and Stability AI. It also provides exclusive access to Amazon's Titan family of foundation models (Figure 1-6). A single model cannot achieve everything. Amazon Bedrock provides a selection of foundation models from prominent vendors, giving AWS customers the option to utilize the best models for their individual needs.

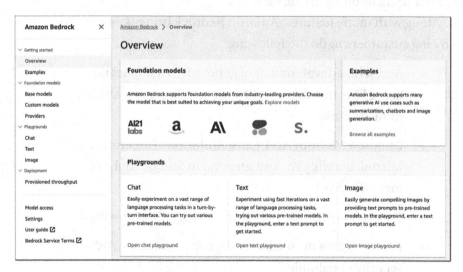

Figure 1-6. *Amazon Bedrock Overview panel*

Amazon Bedrock integrates with Amazon CloudWatch and AWS CloudTrail to meet customers' monitoring and governance requirements. CloudWatch allows tracking of usage metrics and building of customizable dashboards for auditing reasons. CloudTrail allows monitoring of API activity and fixing of issues while integrating other systems into generative AI applications.

Amazon Bedrock can also help to create GDPR-compliant apps and operate sensitive workloads governed by the United States Health Insurance Portability and Accountability Act (HIPAA).

In Amazon Bedrock, the user can browse FMs and load sample use cases and prompts for each model. Once model access is enabled, users can experiment with multiple models and inference configuration parameters to discover one that suits their needs.

Amazon, for example, allows construction of a contract entity extraction use case utilizing Cohere's Command model.

Figure 1-7 depicts the prompt utilized, the inference configuration parameter settings on the right side, and the API request beneath the example answer. If you choose Open in Playground, you can learn more about the model and use case through an interactive console experience.

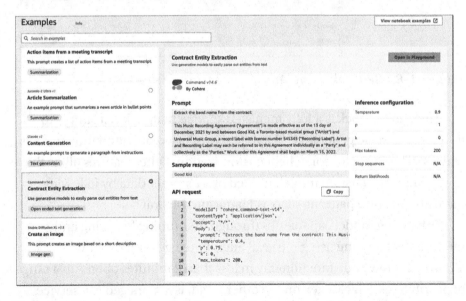

Figure 1-7. Amazon Bedrock using Cohere's model

CHAPTER 1   INTRODUCTION TO ARTIFICIAL INTELLIGENCE

Amazon Bedrock provides chat, text, and picture model playgrounds. In the chat playground, you can experiment with different FMs using a conversational chat interface. Figure 1-8 depicts an example of Bedrock utilizing Anthropic's Claude model.

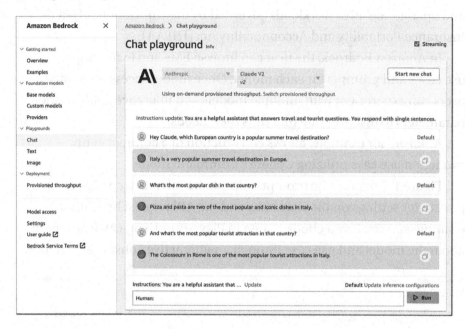

Figure 1-8. Bedrock using Anthropic's Claude model

Amazon Bedrock is a fully managed service that makes it simple to develop and scale generative AI applications based on foundation models while still allowing for some light customization. Other features allow the user to import a custom model, trained on customer data by following main architecture patterns such as Llama and Mistral.

Enhancing model capabilities through fine-tuning featuring and information sourcing using retrieval augmented generation (RAG) allowing for low code functionality in less than 2 minutes opens up a range of possibilities. SageMaker, on the other hand, is a managed ML service designed for more complex use cases, offering extensive customization.

To be more specific, SageMaker trains, builds, and deploys models, whereas Bedrock works with pretrained models. SageMaker allows training of any open-source LLM. In contrast, Bedrock is based on foundation language models, which may be fine-tuned.

Amazon Bedrock provides extensive data security and privacy by storing all data, including prompts, information needed to augment prompts, FM answers, and customized FMs, in the location where the API call is performed. All data is secured in transit with TLS 1.2 and at rest with service-managed AWS Key Management Service (KMS) keys.

AWS PrivateLink can be used with Amazon Bedrock to provide private connectivity between FMs and any on-premises network without exposing user traffic to the Internet. Furthermore, FMs can be adjusted privately, giving complete control of how data is utilized and secured. To do this, Amazon Bedrock makes a copy of the base FM model and trains it offline.

Additionally to safeguard personalized models (FMs), AWS offers security solutions to establish a security plan. Personalized FMs are secured through encryption using AWS KMS keys. With the use of AWS Identity and Access Management (IAM) users have the ability to manage access to their customized FMs granting or denying access to FMs. This feature provides control over which services receive deductions and who has the ability to sign in to the Amazon Bedrock console.

You can supplement organization-specific information to the FM and generate accurate responses using knowledge bases for Amazon Bedrock, which delivers contextual and relevant responses by incorporating organization-specific data through a process known as RAG.

RAG is a technique that involves fetching data from company data sources and enriching the prompt to deliver more relevant and accurate responses and equip the FM with up-to-date proprietary information.

Amazon Bedrocks Knowledge Bases serve as a solution for managing RAG by tailoring FM responses with contextual corporate data in a fully automated workflow from ingestion to retrieval and prompt enhancement without the need for manual coding to connect data sources and handle queries.

Apps are designed to handle conversations by managing session context in order to enable multiturn interactions smoothly. When retrieving information from knowledge bases in apps and systems built on AI technology, it is important to include source citations for transparency and to reduce the likelihood of generating outcomes or hallucinations. These errors can occur due to factors such as training data or biased assumptions made by the model during training using datasets. Linkage of specific data sources can provide field marshals (FMs) with in-depth knowledge about a particular area or organization.

The knowledge bases can access the data residing in Amazon Simple Storage Service (Amazon S3) and convert that data to vectors using embedding models. The vectors can be stored in vector databases such as Amazon OpenSearch Serverless or third-party databases. Agents[3] for Amazon Bedrock can be used to retrieve the relevant documents for a query and deliver accurate responses. Responses from the knowledge base include source attribution, again to increase transparency and minimize hallucinations.

To expand the possibilities of Amazon Knowledge Bases sourcing, Amazon Bedrock has increased its capabilities with native connectors launched during the 2024 NYC AWS Summit, which allow direct connection to custom web sources or popular data sources like Salesforce, Confluence, and SharePoint. This brings more contextual metadata, which drives up the accuracy and drives down the rate of hallucinations inside generative AI applications.

Amazon Bedrock also allows the customer to fine-tune a foundation model with its capability to support a custom-labeled training dataset that improves the model's performance on specific tasks. To fine-tune a model, a training and validation dataset can be uploaded to Amazon S3 to provide a S3 bucket path to a user's Amazon Bedrock job. The Amazon Bedrock console or API can be used for this fine-tuning.

---

[3] Note: Agents is a future trend, as it allows the execution of actions and interpretation and generation of code, which makes it even possible to generate Python code for graphical representation or create an entire app from scratch.

It's important to note that in the case of prompt-based learning and RAG, you are not customizing the FMs. However, when you fine-tune an FM, you are customizing the FM by creating a private copy of the FM.

Amazon Bedrock has been used to support many typical generative AI use cases:

Text generation: Generates content, like narratives and articles, for social media entries and website content

Chatbots: Develops chatbots and virtual assistants to enhance user experiences for customers

Search: Utilizes an amount of data to find and combine information to address questions

Text summarization: Gets an overview of written content, like articles or blog entries to grasp the points without going through the entire piece word by word

Image generation: Creates creative portrayals of topics and environments by leveraging linguistic cues

Personalization: Assists customers in finding exactly what they're looking for by offering targeted and meaningful product recommendations based on context and relevance

Images and document analysis: Analyzing images and documents to enhance capabilities

CHAPTER 1   INTRODUCTION TO ARTIFICIAL INTELLIGENCE

Amazon Q Developer

Picture yourself as a software developer working alongside an AI-powered coding assistant to streamline and simplify the coding process—that's what Amazon Q Developer offers.

As of April 30, 2024, Amazon CodeWhisperer is now Amazon Q Developer, the artificial intelligence coding companion capable of generating real-time, single-line, or full-function code suggestions. It also can explain code (Figure 1-9) in your integrated development environment (IDE) to help build software more quickly. It also supports CLI completions and natural-language-to-bash translation in the command line.

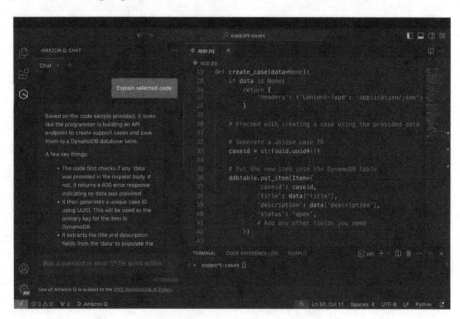

Figure 1-9. Amazon Q Developer code explained

It employs AI behind the scenes to offer coding recommendations by analyzing a user's feedback and their previous code snippets. Individual developers have the option to utilize a version of Amazon Q Developer or opt for a paid version tailored for purposes that includes enhancements such as enterprise-grade security and administrative functions.

The Q Developer programming language supports Python, Java, JavaScript, TypeScript, and C#. Q Developer is also capable of generating code suggestions for Go, Rust, Kotlin, Scala, Ruby, PHP, SQL, C, C++, and Shell Scripting.

The experience of developing code with Amazon AI capabilities can be supercharged even more with the expert assistant Amazon Q.

Amazon Q, an interactive, generative AI-powered assistant available in the IDE through Q Developer, provides professional advice via a simple conversational interface. Some of the capabilities highlighted are as follows:

> **Explain code**: Initiate an interaction with Amazon Q to help comprehend project code using natural discourse (Figure 1-10).
>
> **Transform the code**: Upgrade and migrate your application to the most recent language version in minutes.
>
> **Receive tailored code suggestions**: Ask Amazon Q to give recommendations for implementing unit tests, troubleshooting, optimizing code, and more.

CHAPTER 1   INTRODUCTION TO ARTIFICIAL INTELLIGENCE

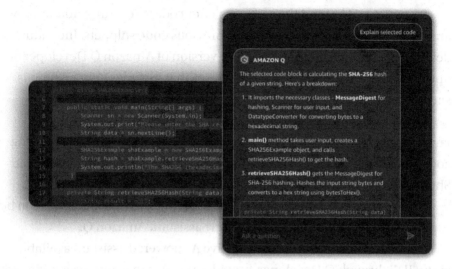

Figure 1-10. *Amazon Q Developer*

Amazon EC2 Inf2 Instances Powered by AWS Inferentia2 Chips

The handling of ultra-large ML models requires massive computational power to run them. AWS introduced the Inferentia chips to offer the most energy efficiency with the lowest cost for running demanding generative AI inference workloads (often involving running models and responding to queries in production) at scale on AWS.

New Trn1n Instances, Powered by AWS Trainium Chips

Generating AI models must undergo training to provide responses such as answers or images to the task at hand. This process is crucial for their effectiveness and efficiency in handling tasks effectively and in a resource-friendly manner. Amazon recently unveiled the Trn1n instance, which is a server resource powered by AWS's custom Trainium chip that facilitates operations efficiently by leveraging networking capabilities essential for swift and cost-effective model training.

SAP Joule

SAP unveiled Joule in September 2023, presenting it as a breakthrough natural-language, generative AI "copilot" that will alter the way enterprises operate.

Joule was created to be an advanced natural-language, generative AI assistant—a trusted AI companion by your side, ready to support and improve your everyday operations by using the power of natural language and generative capabilities throughout SAP's cloud portfolio. It is capable of understanding, analyzing, and processing massive volumes of data based on LLM technology.

By streamlining tasks and offering guidance and assistance to users, Joule enhances user satisfaction and empowers businesses to meet their goals with ease and efficiency. The platform is tailored for integration, with SAP's cloud-based business solutions delivering personalized recommendations sourced from a wide array of SAP applications and external data providers.

The Joule copilot aims to assist people in completing tasks and achieving improved business results securely and compliantly by quickly organizing and interpreting data, from various systems to unveil valuable insights.

Users may simply ask Joule a question or pose an issue in plain language to obtain intelligent responses based on the abundance of business data available throughout the SAP portfolio and third-party sources while maintaining context (Figure 1-11).

CHAPTER 1 INTRODUCTION TO ARTIFICIAL INTELLIGENCE

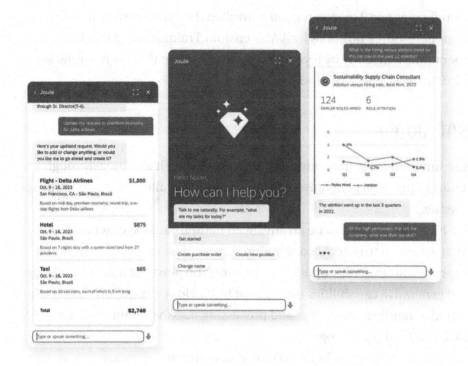

Figure 1-11. *Joule in action*

Consider the example of a manufacturer asking Joule for aid in assessing sales performance. Joule may detect underperforming regions, link to other data sets that highlight a supply chain issue, and instantly connect to the supply chain system to propose viable solutions for the company to consider. Alternatively applied in the human resources, it may aid in the creation of objective job descriptions and pertinent interview questions.

Joule is neither a substitute nor a successor to SAP Conversational AI, a discontinued platform for creating personalized chatbots and allowing users to develop their own content. The copilot has been seamlessly integrated into SAP systems, delivering immediate benefits.

CHAPTER 1 INTRODUCTION TO ARTIFICIAL INTELLIGENCE

Joule offers a conversational user interface that integrates with SAP applications. It is a feature-rich web client that presents assistant replies with SAP Fiori-compliant UI components.

Joule offers out-of-the-box interaction with SAP back-end systems, and it complies with AI ethics, GDPR, and privacy rules while being SOC-II compliant. It is automatically updated as new capabilities are added or altered.

Joule is also connected with SAP Start. It is accessible with SAP S/4HANA Cloud's public version. After that, it will be available in SAP Customer Experience, SAP Ariba, and SAP BTP. The Joule copilot has also been integrated into SAP Business Application Studio.

SAP Build Code

In November 2023, SAP released SAP Build Code, a solution meant to simplify the whole lifetime of corporate applications. SAP Build Code provides a turnkey environment for developing, testing, integrating, and maintaining applications by combining important design time, runtime services, and tools into a single solution.

The SAP Build Code environment combines SAP Business Application Studio's development environment with the most important services and software development kits (SDKs) on the SAP Business Technology Platform (BTP). The goal is to drastically improve the development experience, increase efficiency, and supercharge production. It enables developers to build quickly utilizing AI code generation with Joule copilot, which includes developing application logic from natural language descriptions.

Upon starting a new project from the Build Code lobby, the Business Application Studio often launches with the Storyboard viewer open. To help developers use generative AI capabilities more efficiently inside Business Application Studio, a new tutorial titled "Generative AI-Powered Development" has been published, which can be found in Joule.

CHAPTER 1   INTRODUCTION TO ARTIFICIAL INTELLIGENCE

In addition to the standard way of manually creating the data model with the Core Data Services (CDS) graphical modeler tool in Business Application Studio, Joule can be requested to construct it using a pure natural language prompt (Figure 1-12).

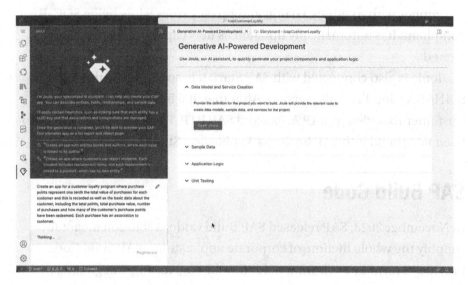

Figure 1-12. *Using Joule in SAP Build Code*

Typically, the answer that completely meets the requirements will be delivered, and it may be accepted immediately into the working project by hitting the Accept button in the top-right corner of the code block inside the response content (Figure 1-13). The storyboard will be updated automatically to reflect the freshly generated results.

As you can see, Joule in Business Application Studio with SAP Build Code not only generates code but also supports SAP Business Application Studio's visual modelers, allowing customers to continue to utilize the current visual editors.

CHAPTER 1   INTRODUCTION TO ARTIFICIAL INTELLIGENCE

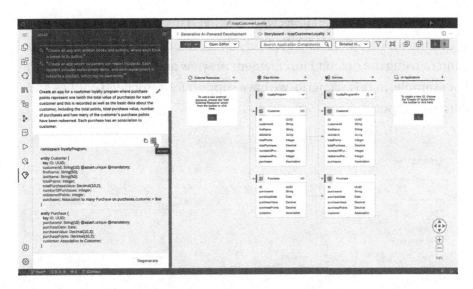

Figure 1-13. *Code generation with Joule*

In addition to the creation of the data model and data service, contextual sample data (Figure 1-14) is established in response to the user's query.

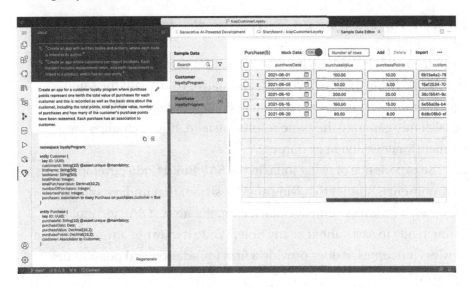

Figure 1-14. *Sample data generated by Joule*

47

CHAPTER 1   INTRODUCTION TO ARTIFICIAL INTELLIGENCE

In addition, UI annotations were created. This indicates that the application is complete, and the user can run a brief preview of the created content using the default Fiori Element preview application (Figure 1-15). For example, in this scenario, we have support for associated entities and date-type inputs.

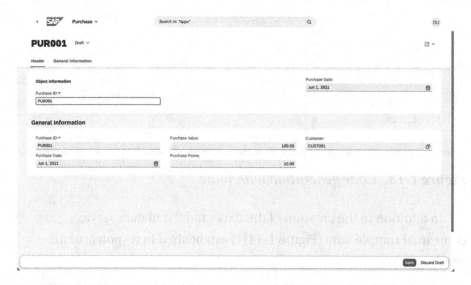

Figure 1-15.   Previewing a Fiori app created by Joule

If the result is inadequate to the user, they can request that Joule regenerate the response to view an alternative result with the same user prompt or a narrowed request. Keep in mind that Joule does not consistently generate the same results. Instead, minor changes are frequently generated in its output.

The experience of using Joule in the SAP Build Code development environment is similar to having an AI assistant that understands a customer's business context while having the newest SAP technology sitting ready to assist them at any time. While it may not always produce flawless outcomes, it does provide a firm foundation and possibilities for growth.

CHAPTER 1 INTRODUCTION TO ARTIFICIAL INTELLIGENCE

SAP AI Foundation

SAP has launched *AI Foundation*, a one-stop shop for developers to create AI- and generative AI–powered extensions and applications for SAP Business Technology Platform (SAP BTP).

AI Foundation is SAP's all-in-one AI toolkit (Figure 1-16), providing developers with AI that is ready to use, customizable, based on business data, and backed by leading generative AI foundation models. It also serves as the foundation for SAP's AI capabilities, which are integrated throughout its portfolio.

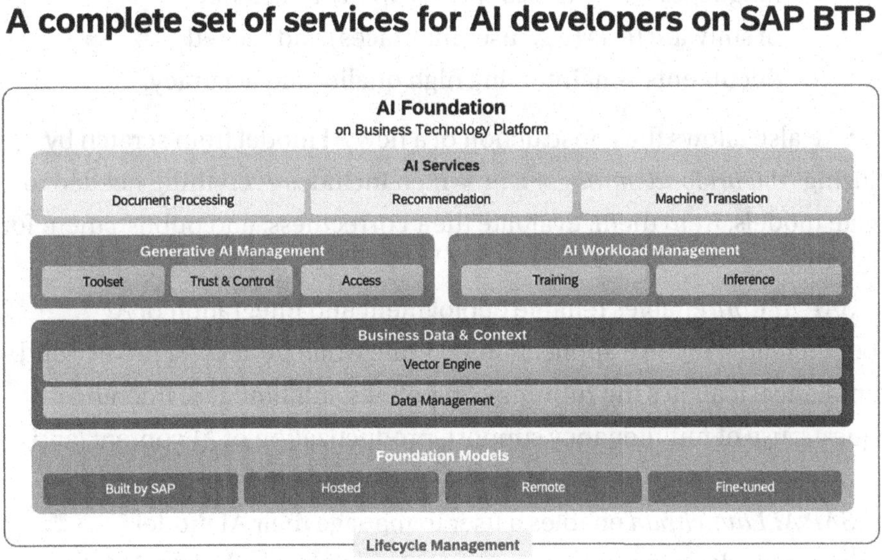

A complete set of services for AI developers on SAP BTP

AI Foundation
on Business Technology Platform

- AI Services: Document Processing | Recommendation | Machine Translation
- Generative AI Management: Toolset | Trust & Control | Access
- AI Workload Management: Training | Inference
- Business Data & Context: Vector Engine | Data Management
- Foundation Models: Built by SAP | Hosted | Remote | Fine-tuned
- Lifecycle Management

Figure 1-16. AI Foundation

AI services are AI models that have been trained using data to business to improve applications for customers catering to business needs and scenarios.

- **Document processing**: This service enables automated document processing that reduces human labor and expenses through information extraction. This enables enormous volumes of business documents to be processed with material in headers and tables.

- **Recommendations**: This service enables personalized and data-driven to improve decision-making.

- **Machine translation**: SAP Translation Hub is used for machine translation of information into several languages. This service speeds up the translation of software texts (*e.g.*, user interfaces) and related documents—maintaining high quality and accuracy.

SAP also allows the construction of a new AI model from scratch by utilizing *AI workload management*, which includes everything needed to create models, train them, evaluate their correctness, and publish them for inference.

SAP AI Core enables reliable deployment and integration of AI models created for SAP applications at scale while protecting privacy and compliance. It allows the running of pipelines, fulfillment of inference requests, use of multitenancy support, productization of AI content, and use as a service to customers in the SAP BTP marketplace.

SAP AI Launchpad enables a user to manage their AI models transparently. It connects to numerous AI runtimes, including SAP AI Core, and centralizes AI lifecycle management for AI scenarios input through a simple user interface. You may continually examine the model's performance data and retrain as needed.

SAP provides *generative AI management*, which functions like a *generative AI hub*, to help accelerate a user's generative AI development. It offers rapid access to a wide selection of LLMs from several suppliers, including GPT-4 by Azure OpenAI and OpenSource Falcon-40b.

This access allows for the coordination of numerous models, either programmatically using SAP AI Core or through the playground within SAP AI Launchpad.

Generative AI management enables tooling for prompt engineering, experimentation, and other capabilities to speed up the building of SAP BTP applications that include generative AI in a safe and trustworthy manner. Using it, AI development teams may send a prompt to numerous LLMs, compare the results to determine the best model for the task, and get more control and transparency with the built-in prompt history.

SAP HANA Cloud enables the integration of AI with business data and context through the vector engine and similarity search functions. SAP HANA Cloud's AI function libraries (Predictive Analysis Library and Automated Predictive Library) enable the application of classification, regression, or time-series forecasting scenarios applied directly to business data, which can be orchestrated using SAP AI Launchpad.

Furthermore, the new generative AI hub in SAP AI Core has been designed to connect to the new *vector capabilities in SAP HANA Cloud* allowing developers to reduce model hallucinations and incorporate contextual data as embeddings (or groundings) to deliver more tailored results to specific use cases.

A *vector datastore* stores unstructured data (such as text, photos, or audio) in high-dimensional vectors or embeddings to offer long-term memory and context to AI models. This allows for rapid identification and retrieval of related things by asking a query in natural language. This streamlines interactions with LLMs and enables developers to safely apply generative AI in apps.

SAP HANA Cloud is designed to natively store and explore vector embeddings, or numerical representations of objects, as well as business data, as part of its industry-leading multimodel processing capabilities to enable intelligent data applications. SAP HANA Cloud's vector capabilities will enable RAG, allowing the merging of LLMs with private corporate data.

To help with data management, SAP has positioned the solution *SAP Datasphere*, a multicloud data fabric solution that enables both hybrid and cloud-native architectures. It leverages AI by allowing the integration of data with AI platforms to deliver data in business terms from multiple systems and locations.

SAP Datasphere uses AI in a variety of ways, including data integration, data quality, cleansing (*e.g.*, restoring missing values or duplication), data governance, analytics (*e.g.*, recognizing patterns, analyzing historical data, and making predictions), and natural language processing.

SAP collaborates with top AI and LLM vendors to enable customers the necessary flexibility to keep up with rapid innovation.

To maximize value in traditional business domains like finance, sales, and supply chain, customers need business-specific foundation models. SAP's approach is to fine-tune generic big language models using SAP anonymized data while also developing proprietary foundation models based on enormous, structured business data (Figure 1-17). These models will be able to answer business-related issues that huge language models cannot, such as forecasting invoice payment deadlines and supplier delivery quality, as well as recommending efficiency enhancements to a business process.

CHAPTER 1   INTRODUCTION TO ARTIFICIAL INTELLIGENCE

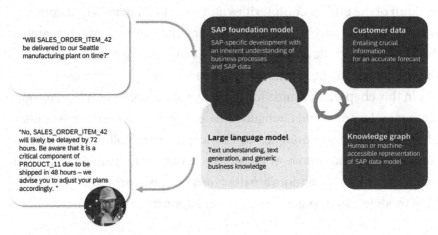

Figure 1-17. Foundation models designed for business context

The Future of AI

Recently, in the AI field there have been advancements in AGIs, AI technology, and LLMs. These developments show potential in making AI systems more intelligent and creative like humans are perceived to be capable of being. The emergence of cutting-edge tools like the GPT series and ChatGPT signifies a phase in AI research that aims to create more adaptable AI systems capable of adjusting to different scenarios and environments.

Despite the progress made, there are still significant challenges to overcome in these subdomains, particularly when it comes to ensuring the quality and authenticity of generated content and developing ethical guidelines for creating and using AI systems. However, the potential benefits of these technologies are enormous, ranging from new forms of creativity and expression to improved healthcare, scientific discovery, and engineering innovation.

CHAPTER 1   INTRODUCTION TO ARTIFICIAL INTELLIGENCE

With the impact of AI, in our society today and its evolving role in shaping our landscapes, it is crucial for us to keep a close eye and be mindful of the ethical complexities and potential hazards that come with its creation and implementation. Collaboration among researchers developers and policymakers plays a role in steering these advancements toward enhancing the welfare and success of everyone involved.

In this chapter, I've introduced many aspects of AI as it began, gained stronger relevance, and changed in recent years, having covered many of its important terminologies and concepts. In the following chapter, we will go in depth into how disruptive AI has been during and after the pandemic. Most IT vendors are now on a fast pace developing their own LLM models and AI platforms to better support different use cases.

CHAPTER 2

GenAI in the Spotlight

In the realm of intelligence (AI) there have been developments in three key areas: artificial general intelligence (AGI), generative AI (GenAI), and large language models (LLMs). These domains have been instrumental in driving advancements in the field of AI.

AGIs aim to develop machines that can learn and adapt like humans do; generative AIs are programmed to generate content independently, demonstrating creative abilities with little human input.

Despite their different goals, both subdomains of AI are driven by the broader goal of creating machines that can demonstrate more autonomy and creativity in their decision-making processes. In fact, some experts believe that the development of generative AI is a crucial step toward achieving AGI, as machines that can generate their own content and ideas are more likely to learn and adapt to match human expectations.

Reshaping Industries with Generative AI

GenAI shows great promise to revolutionize various industries and fields.

Using machine learning algorithms to produce content like images and music that closely resemble the style and quality of the input data is a common practice in this field. By examining many data sets thoroughly, these algorithms have the ability to create original and distinct content that brings about new avenues for creativity and self-expression. For instance, an AI system that has been taught using images of faces has the capability to generate realistic facial images that have never been seen before.

Various applications have already used generative AI for anything from visual art and music composition to literature and video production. A growing number of platforms and tools are available that make it more accessible to developers, artists, and creatives.

There has been progress in LLMs particularly in the realm of artificial intelligence that allows them to understand and process languages more efficiently. These models fall under the umbrella of natural language processing, which deals with the ways computers can grasp and interpret languages effectively. LLMs leverage methods, like machine learning and deep neural networks, to analyze and produce language.

The combination of GenAI, AGI, and LLMS has the potential to create powerful and innovative applications in various domains.

According to Precedence Research, the overall value of the AI market is projected to hit $150+ billion in 2023, rising to as much as $1.5 trillion by 2030 (Figure 2-1).

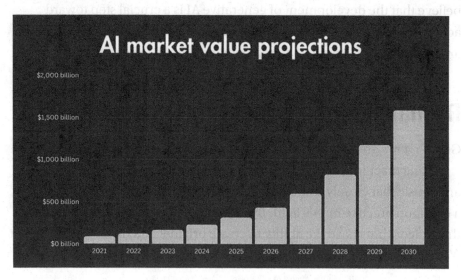

Figure 2-1. *Artificial intelligence market size 2021–2030*

CHAPTER 2   GENAI IN THE SPOTLIGHT

The Pandemic and AI Disruptions

In the past three years there have been remarkable strides and changes in the realm of artificial intelligence leading to notable shifts across various industries worldwide. The year 2020 was marked by the onset of the COVID-19 pandemic, which hastened the integration of AI solutions in fields like healthcare and manufacturing. Despite these hurdles, AI has significantly reshaped our business landscape. Breakthroughs in areas such as natural language processing, computer vision, and reinforcement learning have paved the way for AI tools and applications to emerge.

The healthcare crisis caused by the pandemic led to a rise in the need for AI-driven solutions in the medical sector. Experts and healthcare workers swiftly innovated AI tools to aid in the diagnosis and treatment of COVID-19. For instance, they developed AI algorithms to spot the signs of the virus in chest X-rays, CT scans, and other medical images, with increased speed and accuracy. These algorithms could pinpoint characteristics and patterns in the images that indicated a presence, enabling swifter detection and intervention strategies. The advancement of these AI models serves as an illustration of how technology is reshaping the healthcare sector and holds the promise of changing our approach to medicine profoundly.

Chatbots and virtual assistants were also used in healthcare settings to assist providers in handling patient queries and offering advice. They could respond to questions and could offer details on symptoms, test locations, and treatment choices. Specific chatbots could even evaluate patients for COVID-19 and assess their risk level depending on their symptoms and medical background.

During this period, AI technologies were not utilized only in healthcare but also in the field of education. AI-driven platforms in education enabled learning and offered tailored feedback to students. Additionally, AI chatbots and virtual assistants were utilized to offer health assistance and counseling to individuals grappling with induced stress and anxiety. These technology-driven solutions played a role in lessening the effects of stress during that time period, and their influence persists even now that the pandemic is over.

The global impact of the COVID-19 pandemic has been profound. It has hastened the integration and growth of AI technologies across sectors significantly. This has led to the emergence of tools and solutions that continue to revolutionize industries and communities. The period from 2021 to 2023 witnessed advancements in cutting-edge AI technologies, like AI models, vast language models, and artificial general intelligence systems.

At the same time, there has been a growing trend toward the democratization of AI, making it more accessible to individuals and small businesses. AI platforms and tools have become more user-friendly and affordable, enabling individuals and small businesses to develop and deploy their own AI applications. For instance, many cloud-based AI platforms now offer prebuilt models that can be easily integrated into new applications without requiring extensive technical expertise. This democratization of AI is expected to have a significant impact on the future of innovation, as it allows more diverse voices and perspectives to contribute to the development of new AI applications and tools. This also represents a great opportunity to make different open source models available as the path forward to attend multiple scenarios.[1]

The advancement of deep learning structures have brought about enhancements in fields such as natural language processing and image and speech recognition, among others. These structures have contributed to the creation of more lifelike images and videos as well as enhancing machine translation capabilities and facilitating the development of sophisticated conversational interfaces and chatbots.

---

[1] At the time of this book release, Meta has released Llama 3.1 405B, the first frontier-level, open-source AI model, as well as new and improved Llama 3.1 70B and 8B models. In addition to significantly better cost/performance relative to closed models, the open 405B model intends to be the best choice for fine-tuning and distilling smaller models.

CHAPTER 2   GENAI IN THE SPOTLIGHT

AI has seen an advancement with the rise of quantum computing technology over years. Experts have been making progress in crafting algorithms and constructing quantum computers able to handle more intricate operations. The emergence of quantum computing holds the promise of transforming AI by enhancing the capabilities of machine learning algorithms and expediting processing speeds. For example, quantum machine learning has the potential to streamline the processing and analysis of datasets like those produced by sensors and Internet-connected devices in making immediate decisions and quicker forecasts.

Moreover, the ethical aspects related to AI application have gained significance. Authorities and institutions have laid down standards and rules to guarantee that AI is created and utilized in a moral manner. These regulations aim to stop the abuse of AI and ensure its utilization for the well-being of the community. From 2021 to 2023 considerable endeavors have been made in crafting frameworks and resources to pinpoint and reduce any hazards linked to AI.

Unleashing AI as Never Before

An amazing event happened on a bright but otherwise regular Tuesday, on March 14, 2023, when the fourth version of GPT was released. GPT-4 had an impressive ability to select and use tools to achieve its goals. With its release, a eureka moment happened, mirroring Steve Jobs' famous insight about the power of computers: "They function as bicycles for our thoughts, propelling us to unprecedented heights."

Throughout history, tools have accelerated human advancement. The plow transformed agriculture, and the printing press democratized information, eventually leading to humanity's greatest accomplishments, such as landing on the moon. Now, a new chapter has begun in our story.

CHAPTER 2   GENAI IN THE SPOTLIGHT

March 14 began the era where AI is capable of utilizing tools effectively opening up a realm of possibilities. Soon breakthroughs will challenge the boundaries of human capacity. As we delve deeper into unlocking and leveraging the power of AI, the horizon is brimming with advancements and promising prospects for the future.

Within a week of this milestone, on March 22, Microsoft published the research paper "Sparks of Artificial General Intelligence: Early Experiments with GPT-4," which documented early tool usage discoveries. One day later, OpenAI unveiled plugins, an open-source toolkit that teaches AI how to interact with websites and data, and within a month, open-source projects such as AutoGPT and HuggingGPT dominated GitHub, achieving tasks previously thought impossible.

To be clear, the day AI gained the ability to independently decide which tools to use to solve assignments and learn to acquire new tools will be remembered as one of the most impactful events of this era. Soon, every company will be an AI company and every agency an AI agency. AI will become the new reality, pushing the boundaries of what is possible and elevating our experiences.

In this chapter, we'll look at the broad implications of generative AI for industries ranging from manufacturing to healthcare to banking to transportation and beyond. Together, we'll discover the huge prospects presented by AI while remaining optimistic about how it might transform our world for the better.

As we embark upon an era of advancements and innovations, there is a realm brimming with novel progress and possibilities awaiting exploration.

Healthcare and Biotechnology

Imagine a world where healthcare experiences a drastic transition, rendering modern medicine obsolete as medieval procedures. We are on the verge of a healthcare and biotechnology revolution driven by AI's incredible potential.

Personalized Medicine

AI-driven diagnoses and treatment strategies can advance personalized medicine by responding to each patient's needs. By examining patient data, AI can identify patterns and correlations that lead to more targeted and effective care.

As AI gets more personal and emotionally sophisticated, we should expect a revolution in telemedicine. Consider an ever-present, AI-powered general doctor at your disposal, capable of prescribing tailored treatments for your specific needs.

Dr. Isaac Kohane, a distinguished physician and computer scientist at Harvard, highlighted GPT-4's extraordinary talents in his book *The AI Revolution in Medicine.* He shared that this AI system can correctly answer US medical test license questions more than 90% of the time. In one example, it discovered a rare disease in an infant, which surprised and concerned the doctor.

Medical Imaging Techniques

The integration of AI and tools is revolutionizing imaging enhancing disease detection and categorization, which not only minimizes human error but also accelerates the diagnosis process.

Medication Development

AI technology plays a part in the process of discovering and developing medications by effectively expediting research and pinpointing potential treatment options. Biotechnology firms leverage AI to explore compounds and hasten the progress of drug development through the utilization of AI-created molecules.

DeepMind's AlphaFold has the ability to forecast the configuration of all recognized proteins, which leads to innovative treatment possibilities being available now. Additionally, artificial intelligence accelerates trials by utilizing simulations.

Through the contributions of billionaires and advancements, in technology we are embarking upon an era in healthcare where the remarkable potential of AI is poised to revolutionize both healthcare and biotechnology offering promise for a future for generations to come.

Finance and Investment

The finance and investment industries are undergoing changes as GenAI steps in to revolutionize areas of the markets such as trading strategies and fraud prevention while also automating processes effectively. This shift is expected to create a more effective landscape that will ultimately benefit both professionals in the field and end consumers.

Financial Advice

The use of AI in trading is changing the way portfolio management and risk evaluation are handled in institutions and reshaping how investment strategies are developed.

Improved trading algorithms and the integration of AI technology have empowered experts to make decisions rapidly in response to market changes and tailor investment strategies to meet specific requirements effectively. AI-driven portfolio management and personalized investment strategies can mitigate risks to ensure returns for investors result in increased benefits for investors.

Fraud Management

Fraud detection and prevention are crucial for maintaining financial systems, and they depend on AI-driven systems that continuously monitor transactions and behavior trends to enhance cybersecurity measures by identifying threats and vulnerabilities more accurately while reducing false alarms. This capability also allows for responses to cyber threats.

Financial Services

AI-driven chatbots and virtual assistants are revolutionizing customer interactions by serving as managers that enhance satisfaction through effective customer support services. AI-driven automation streamlines resource management and reduces expenses by automating routine tasks and decision making processes. In essence, AI streamlines services enabling both companies and individuals to concentrate on what matters.

Embracing the AI advancements in the banking and investment sectors could lead us toward a safer financial landscape with increased efficiency.

Information Technology

AI is a key player in the field of information technology. The transformative effects of GenAI are being felt across sectors such as software development, silicon chip manufacturing, data storage facilities, cybersecurity measures, and online search platforms. Let's delve into the influence of GenAI, within these areas.

Software Development

Within software development, AI dramatically improves the processes of coding, debugging, and collaboration. AI-assisted coding and debugging technologies improve code quality, minimize errors, and speed up development cycles.

GenAI-powered collaboration and project management solutions improve communication and coordination, resulting in more efficient software development projects and faster time to market for new products.

Semiconductors and Hardware Design

AI-driven chip design and optimization strategies shorten the design cycle and reduce development costs in the semiconductor and hardware industries. As AI grows more competent at using tools to develop high-performing and energy-efficient chips, the semiconductor sector will see fast growth. AI-powered material discovery reveals the potential for next-generation technology, reduces the time to market for disruptive innovations, and keeps the industry at the forefront of innovation.

AWS and NVIDIA have started a revolutionary collaboration named Ceiba Project that aims to push the boundaries of AI by developing the largest AI supercomputer in the cloud. Hosted exclusively on AWS, this cutting-edge supercomputer will power NVIDIA's AI research and development efforts.

Data Centers and Cloud Infrastructure

AI-enabled energy management and resource allocation solutions are transforming operations in data centers and cloud infrastructure. AI-driven solutions maximize energy efficiency while lowering expenses, providing a more sustainable approach to managing these critical resources. Furthermore, AI-driven maintenance and problem-detection systems reduce downtime and improve service reliability, providing a firm basis for today's digital services.

Cybersecurity

AI-powered threat detection and prevention systems are changing the landscape by identifying and reducing cyber hazards in real time. Systems powered by AI provide strong protection against malicious actors by reacting to constantly changing attack vectors and methods. AI-enabled incident response and recovery solutions accelerate cleanup and limit damage from breaches as they continuously learn and develop to create a more holistic cybersecurity approach.

Search Engines and Information Retrieval

Web 1.0 was static, whereas Web 2.0 was dynamic; nevertheless, Web 3.0 will be conversational. Customers can anticipate engaging chats with websites via an AI interface housed on-site and working as an API. This configuration will enable a generative model to retrieve bespoke data and produce hyperpersonalized content.

AI-powered content generation and curation systems will broaden and deepen the available knowledge base, making it easier to discover and consume. Search algorithms and customization approaches powered by AI will improve search relevancy and user satisfaction by adjusting to user preferences and providing bespoke results. As a result, the huge ocean of knowledge will become more navigable and valuable than before.

Communication Services

In this section, we'll look at how generative AI can transform education, media, music, and the legal sector. We will look at how AI can help us reimagine our way of learning, accessing information, and interacting within these communication services, opening up new avenues for innovation and growth in them.

Education and Learning Experiences

On March 14, 2023, the world witnessed the release of Khanmigo, an AI instructor for Kahn Academy, as well as a new subscription tier for Duolingo to enhance language learning experiences.

As a child, my curiosity often drove me to ask a lot of questions, much to the dismay of my teachers. However, this attitude was critical to my learning success. Now, with GenAI taking on the role of an untiring instructor, the educational environment is on the verge of a momentous transformation, accommodating each learner's unique demands and expediting the learning process.

Discovering facts through AI-powered learning experiences provides a far deeper, more meaningful understanding than passively acquiring knowledge.

These personalized learning experiences can evaluate students' performance and learning styles and then deliver tailored content to improve outcomes. The use of AI-assisted language translation and speech recognition systems removes obstacles to global education, allowing students from all backgrounds to access high-quality educational resources in their native language. This paves the way for a more inclusive and linked learning environment that recognizes the potential of each student.

Media Consumption and Creation

Since December 14, 2022, an AI-generated Twitch channel has been broadcasting a never-ending *Seinfeld* episode. This development reveals the possibility for limitless material from beloved brands, as well as the possibilities for open-world games using AI-powered NPCs (Non Player Character), including those in the metaverse.

AI-powered content production, curation, and recommendation engines are revolutionizing the media industry. These advanced technologies assess user choices, interests, and behaviors to provide tailored information that keeps audiences engaged and informed. Furthermore, AI-powered technologies expedite content creation, allowing for faster and more efficient generation of articles, films, and other media formats. This improves media output quality while reducing the time and resources required to create captivating content.

Music

On March 30, 2023, AllttA dropped a song named "Savages" featuring Jay-Z, but with a twist: AI-generated Jay-Z vocals. The song went viral, showcasing AI's recent advancements. According to YouTube commentator Sam Cornwell, "I think this might be the tipping point that wakes."

AI is transforming music creativity, production, and distribution with AI-powered music composition and production tools. These tools enable musicians to create distinct and original sounds, pushing the limits of creativity.

Legal

Today, AI already excels at bar tests and legal document creation, signaling a significant shift in the field. Artificial intelligence–powered legal research tools and document analysis systems improve the legal profession by automating repetitive work and increasing productivity.

AI-powered solutions help legal professionals make informed judgments and provide precise advice by processing large amounts of data and extracting important insights. Furthermore, AI-powered contract drafting and review technologies reduce errors and ensure legal compliance, saving time and money while improving the quality of legal services for customers.

Digital Advertising

Imagine an AI-powered salesperson capable of tailoring your product or service to each lead, leading them through the sales funnel, and eventually sealing the purchase. This personalized strategy would usher in a new era of advertising, one in which AI targeting and optimization beat traditional ad methods, resulting in a more engaging and efficient conversational experience.

The creative process is already evolving. AI-powered content production solutions like INK, Adobe Firefly, and TypeFace are developing optimized ad copy and images. Simultaneously, systems such as Synthesia and D-ID generate video content that resonates with the intended audience. Businesses can save time and costs by automating these procedures while still providing engaging and compelling commercials.

Furthermore, AI is transforming ad performance assessment and analysis. AI-powered analytics systems provide real-time insights into campaign effectiveness, allowing marketers to make more educated decisions and improve their tactics. Data formerly dismissed as mere noise can now be converted into useful insights via simple discussions with AI systems, reducing the need for specialist knowledge.

AI has become a partner in the fight against ad fraud to improve the transparency and trustworthiness of the advertising industry. Companies can use AI-based tools to identify and combat activities effectively while safeguarding their investments and maintaining the credibility of their advertising efforts.

In essence, the remarkable abilities of AI are leading to a change in the advertising field and setting the stage for more effective marketing strategies. With the rising acceptance and implementation of AI-based methods by companies we anticipate a phase of innovation and expansion in the advertising industry.

Capital Goods

The capital goods industry, which includes businesses such as aerospace and defense, is on the verge of tremendous disruption thanks to AI and its ability to use tools. Let's look at how AI is transforming these industries, as well as its potential applications in other capital goods sectors.

Aerospace

AI-powered innovations are revolutionizing the aviation industry across including development and production, flight control, and the upkeep of aircrafts. Leveraging the design capabilities of AI in tailoring airplane frameworks to enhance aerodynamics and cut down on weightage substantially contributes to fuel economy and environmental sustainability. Additionally, predictive maintenance systems empowered by AI have the ability to identify issues in advance, ensuring safety standards of aircrafts and minimizing downtime effectively.

Defense

The introduction of AI-powered technologies is causing a paradigm shift in the defense sector as well. AI is changing military operations through the use of everything from self-driving drones and enhanced surveillance systems to cyber defense and intelligent logistics. Defense forces can obtain a competitive advantage by using AI for decision-making and strategic planning while limiting human life hazards.

Drone warfare has gained popularity over the last decade with autonomous drones, tanks, and robotics marking the next frontier. We should also expect novel cyber warfare powered by AI advances, both offensively and defensively.

Other Capital Goods

In addition to aerospace and defense, other capital goods industries are being heavily impacted by AI, including construction and heavy machinery. AI-powered construction planning tools can improve project schedules, resource allocation, and cost management, while intelligent heavy machines can accomplish complex jobs with precision and efficiency.

A fascinating amount of potential exists in the intersection between additive manufacturing and swarm construction, both powered by AI. Even the most ambitious building activities will be manageable, and predictive maintenance will lower the cost of running larger projects.

To summarize, the integration of AI and its capacity to use tools is ushering in a new era of innovation and optimization in the capital goods industry. By leveraging the power of AI, these segments may reach higher levels of productivity, safety, and sustainability, resulting in a better future for everyone.

CHAPTER 2   GENAI IN THE SPOTLIGHT

It has become clear that generative AI has enormous transformational potential. This technology is on the point of becoming an important force, reshaping our environment in ways we can't comprehend. AI has the potential to transform every area of our lives, from healthcare and biotechnology to banking, information technology, communication services, and capital goods, allowing for unprecedented levels of efficiency, creativity, and innovation.

Businesses in all industries can harness AI to reinvent their products and services, improve consumer experiences, and adapt to ever-changing market demands.

ChatGPT Spotlights GenAI

The growth of artificial intelligence is one of the most intriguing advancements of the previous decade. With the use of machine learning and deep learning algorithms, AI systems have gotten more sophisticated in their capacity to grasp and process natural language. One of the most impressive instances is the GPT model, which has rapidly gained popularity.

GPT was developed by OpenAI, a research lab founded by multiple tech titans, including Elon Musk and Sam Altman. It was introduced in 2018 and has since become one of the world's most popular AI models.

GPT is a cutting-edge deep learning method that uses a transformer-based architecture to generate natural language responses to prompts. Essentially, it is a machine learning system capable of producing human-like responses to text input.

One of the most important reasons for GPT's success is its adaptability. The model can process and generate text in a number of circumstances, making it suitable for a wide range of applications, including chatbots and virtual assistants, as well as automated content production and translation. As a result, it has the potential to become an indispensable tool for many enterprises and organizations seeking to streamline operations and enhance the customer experience.

However, GPT's influence has extended well beyond the realms of business and technology. It has become a pop culture phenomenon, appearing in a range of mediums, including social media, movies, and television series. One of the most well-known examples of GPT's pop cultural effect is the character Lil Miquela, who is a virtual influencer with a large Instagram following. Lil Miquela is driven by GPT and can generate natural-sounding captions and responses to comments.

Another example of GPT's influence is the recent release of the film *Sunspring*, which was totally written using a GPT algorithm. The film, produced by Oscar Sharp and Ross Goodwin, is a peculiar sci-fi short that demonstrates GPT's immense creative potential. Since its debut, *Sunspring* has become a cult classic, inspiring a new generation of filmmakers to investigate the potential of AI-generated material.

GPT has had a tremendous impact not only on popular culture but also in the realm of artificial intelligence. The model has been the topic of countless academic papers and research efforts, and it has influenced the development of many new AI systems built on similar transformer-based architecture. As a result, GPT has become a key component of modern AI research and development.

A Look at the Evolution of GPT Models

As you have learned, ChatGPT relies on a complex architecture of neural networks and hundreds of billions of parameters to be adjusted iteratively. The principle is simple: for a given sample sentence, predict the next word.

The AI tool thus determines the meaning of a word by taking into account the contexts in which it has encountered it. Training a model with so many parameters requires a large amount of data, generally taken from publicly accessible data such as Wikipedia (3% of the training data), press articles, open-access books, and so on.

As an example, GPT-3, composed of 175 billion parameters, required the ingestion of almost 570 GB of data, or around 300 billion words. These pharaonic figures are synonymous with considerable financial costs—we're talking about 4.6 million dollars spent on training GPT-3[2], not to mention the resulting ecological impact.

The evolution of the GPT model has been observed by individuals, consumers, researchers, and IT companies, who are now paying more attention than ever to the possibilities that enhanced models can offer. The evolution over the last few years can be seen clearly in Figure 2-2.

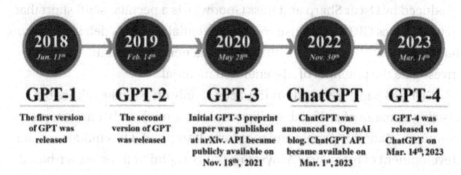

Figure 2-2. *Evolution of GPT models from GPT-1 to GPT-4*

ChatGPT is a chatbot based on GPT capable of emulating the human experience of a real conversation. Thirteen thousand question-answer pairs were used to transform a "next-word prediction" model into one capable of answering questions. In parallel, a reward model—reinforcement learning—helps ChatGPT to orient itself toward producing answers expected by a human. This final stage, during which humans rank different possible answers for the same question, enables the moderation of certain content considered illegal or dangerous.

---

[2] At the time of this book's release, GPT-4o (GPT-4 Omni), a multilingual, multimodal generative pre-trained transformer has been released by OpenAI with capabilities to process and generate text, images, and audio. This application programming interface (API) is twice as fast and half the price of its predecessor, GPT-4 Turbo.

CHAPTER 2 GENAI IN THE SPOTLIGHT

The result has been some other very powerful AI models capable of understanding all the semantics of each language on which they've been trained, enabling them to carry out numerous use cases, such as text translation, essay writing, and so on. More broadly, they can be specialized or adapted to a use case, giving them a wide range of possible tasks, like ChatGPT.

The number of parameters used to train GPT models has increased significantly, as you can see in Figure 2-3.

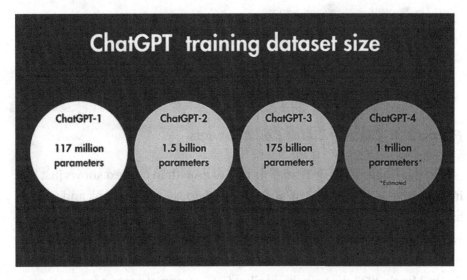

Figure 2-3. *Evolution of data set sizes from GPT-1 to GPT-4 model*

ChatGPT's user base saw expansion compared to online platforms with more than 100 million users flocking to the platform just over two months since its debut—a record-breaking growth rate in consumer software history! The dataset powering Chat GPT 4 clocks in at around 1 trillion parameters—more than five times than what fueled Chat GPT 3s capabilities. Statista reported that Chat GPT attracted a million users within five days of its debut in November 2022 (Figure 2-4).

Figure 2-4. *ChatGPT with 1 million users*

To put Chatbot GPTs incredible rate of growth in context shows just how impressive it is, it took TikTok and other top apps like Facebook and Canva almost 5 years and 9 years, respectively, for their breakthrough moments with users! Instagram also took 2 and a half years to hit the 100 million user mark, an achievement that Chatbot GPT managed in about 60 days.

In March 2024 the website of Chat GPT saw an average of more than 101 million visitors daily.

Ascension of Conversational AI Interfaces

Chatbot GPT has become quite popular in language processing circles for its ability to generate responses that sound like they're coming from a real person when given text prompts in everyday language situations. While it's mainly used for automating customer service tasks and providing support services to users, user interactions have turned Chatbot GPT into an icon influencing social media content, music videos, and more in various ways.

In addition, the popularity of ChatGPT has been greatly boosted by the proliferation of chatbots and virtual assistants in years as businesses strive to enhance customer interactions through effective services provided by these AI powered tools. Chatbots created using Chat GPT can provide businesses with the ability to enhance their customer service operations, effectively enabling them to offer efficient assistance without requiring intervention.

Pop Culture Influence

With the rise of chatbots and virtual assistants, Chat GPT has emerged as a cultural presence. Its impact is evident across forms of media such as articles, news stories, and even the arts.

One clear indication of Chat GPTs impact is evident in media platforms where it has become integral for content creation and engaging with followers. This is due to the surge in social media influencers using it to provide responses that enhance interaction with their audience.

Chatbot influence extends beyond media platforms to various industries, like the music sector where it has been leveraged to craft song lyrics as well as entire albums like *I AM AI*, which debuted in 2021 featuring music solely created with AI technology including chatbots.

The impact of chatbots extends to the gaming industry well. Virtual assistants and chatbots are increasingly prevalent in video games now. Chatbots are even being leveraged to create AI-driven NPCs (player characters) allowing them to engage with players through natural language interactions.

Revolutionizing Linguistic Education

GPT has a talent in aiding language learners by serving as an AI language companion to enhance their proficiency in various languages through interactive question and answer sessions. The GPT technology has also been used to create chatbots that offer personalized feedback and suggestions to individuals learning a language by analyzing their language usage and errors.

AI Trends Shaping Pop Culture

Current culture reflects how today's generation interprets ideas across fields like art and music as well as in literature and science. It's clear AI has caught the attention of pop culture lately with thinkers vying to innovate in the realm; next in line are some key elements to mull over when it comes to any societal trend.

It's interesting to point out that in the past it was not common for the creator of a trend to receive recognition when it became successful; people generally need time to adapt to fresh concepts and integrate them into mainstream culture. It might also require generations of polishing before an idea successfully moves beyond the stages and gains widespread acceptance. By that time newer and established entities are likely to enjoy the rewards of this development.

For example, Google adopted a pattern set by Yahoo. MySpace laid the foundation for Facebook. The balance is maintained by the bell curve. A novel style takes long to become popular as it takes to originate from that culture.

ChatGPT quickly became part of culture in under two months making a significant impact on mainstream pop culture at a rapid pace; however historical trends suggest that its lasting influence will be in line with its swift integration into popular culture realms today—a clear reflection of society's fleeting attention span akin, to that of a mosquitos.

CHAPTER 2   GENAI IN THE SPOTLIGHT

An Exciting Era Unfolds

A week passed after Chat GPT was made available to the public before I stumbled upon it. I got wind of Chat GPT through an episode of *The Daily* podcast by the *New York Times*.

The segment was named "The Battle for Online Searches" focusing on Microsoft's search platform Bing and its new search tool powered by OpenAI's GPT technology with a promise to revolutionize the way countless individuals navigate the Web.

There has been a shift in the way artificial intelligence is utilized on the Internet recently. The media extensively reported on the introduction of chatbot GPT as an AI application that offers a user-friendly platform or a large language model. A multitude of individuals promptly adopted it, and it quickly gained popularity.

However, soon after its creation, it became more of a novelty, something to play around with for entertainment's sake. People would test its capabilities by asking questions and then sharing the answers with others. For instance, they might ask it to craft a tale. It would oblige. Then they'd challenge it to re-create the story in language that resulted in a comical outcome. Following that was a request for a love story narrated by a 1940s gangster. And ChatGPT flawlessly incorporated all the nuances of that era.

It was fascinating to see how the tool came up with all those responses out of nowhere; however, it seemed like such a novelty that people lost interest in quickly, kind of like the excitement surrounding 3D glasses fading away after just a couple of weeks after unboxing them.

Weeks later Microsoft invited a bunch of journalists to discuss a new development in artificial intelligence. They introduced the AI-driven update for Bing. If you're not familiar with Bing, it's Microsoft's search engine that some folks in the tech world consider to be a bit of a laughing stock when compared to search engines.

77

CHAPTER 2   GENAI IN THE SPOTLIGHT

Reimagining Bing: A New Chapter

Microsoft recently revealed its plan to incorporate OpenAI technologies into Bing directly to combine its search engine with OpenAI's advanced language model AI for the revamped Bing platform launch. We should pause to reflect on this significance. This minor technological advancement actually has the potential to push humanity forward into the era of evolution.

Searching has become the cornerstone of interactions and has undergone a significant transformation in recent times. Jumping into a new distinctly innovative search methodology holds significance in our tech savvy society today.

Over the past three decades or more when you look for something using a Google search engine results page (SERP), you'll typically see ads and images. The main focus is a lineup of links to various websites that Google thinks can best address your query or question at hand. It's then up to you to click through these links and sift through their content to find the information you're seeking.

Microsoft aims to use this technology to develop a search protocol that leverages AI to provide personalized responses tailored to each query directly. For instance, I recently conducted a search using the AI-powered Bing search engine.

> *I'm making a 28th wedding celebration dinner for my wife; the main course is grilled salmon, and I need some side dishes to go with it. She doesn't like it too spicy. What should I make?*

> Here is the complete answer I got:

> *For a 28th wedding celebration dinner featuring grilled salmon as the main course, and considering your wife's preference for non-spicy food, here are some elegant yet simple side dishes that will complement the salmon beautifully:*

CHAPTER 2   GENAI IN THE SPOTLIGHT

1. **Lemon Herb Roasted Potatoes**:
   - *Small potatoes roasted with lemon zest, olive oil, garlic, and a mix of herbs like rosemary and thyme. This dish is flavorful without being spicy and pairs wonderfully with salmon.*

2. **Grilled Asparagus**:
   - *Lightly seasoned with olive oil, salt, and pepper, grilled asparagus offers a crispy and slightly smoky flavor that enhances the salmon's taste.*

3. **Garden Salad with Avocado and Balsamic Vinaigrette**:
   - *A fresh salad with mixed greens, slices of avocado, cherry tomatoes, cucumber, and a drizzle of balsamic vinaigrette. It's refreshing and adds a nice textural contrast to the meal.*

4. **Cucumber Dill Salad**:
   - *Thinly sliced cucumbers and red onions tossed in a light dressing made from yogurt, lemon juice, fresh dill, salt, and a touch of honey. This salad is cool and creamy, providing a mild accompaniment to the salmon.*

5. **Quinoa Pilaf with Lemon and Herbs**:
   - *Cooked quinoa mixed with sautéed carrots, celery, and leeks, then flavored with lemon juice and fresh herbs like parsley and chives. It's a nutritious side that adds a hearty element to the meal without overwhelming the main course.*

6. **Steamed Green Beans with Almonds**:
   - *Green beans steamed to perfection, tossed with toasted sliced almonds, and a hint of garlic, then finished with a squeeze of lemon juice for a touch of acidity.*
7. **Wild Rice with Mushrooms and Leeks**:
   - *A savory side that combines wild rice with sautéed mushrooms and leeks, seasoned with a bit of thyme. It's a flavorful but not spicy dish that complements the richness of salmon.*

Remember, the key to a memorable celebration dinner is not just the food but the love and thought put into it. Your wife will surely appreciate the effort you're making to create a special evening for both of you. Enjoy your celebration!"

In addition to its endearing reply, I received a list of groceries for all the ingredients required to prepare the recommended side dishes.

Instead of a bunch of websites offering possible solutions to my query, the updated search tool had almost a human-like conversation with me, providing all the necessary responses I was looking for.

Microsoft Copilot

Microsoft Copilot is a chatbot developed by Microsoft. Launched on February 7, 2023, it is based on a large language model. It can cite sources, create poems, and write songs.

Copilot is Microsoft's primary replacement for the discontinued Cortana. The service was originally introduced under the name *Bing Chat* as a built-in feature for Microsoft Bing and Microsoft Edge. Over the course of 2023, Microsoft began to unify the Copilot branding across its various

CHAPTER 2   GENAI IN THE SPOTLIGHT

chatbot products. At its Build 2023 conference, Microsoft announced its plans to integrate Copilot into Windows 11, allowing users to access it directly through the taskbar. In January 2024, a dedicated Copilot key was announced for Windows keyboards.

Copilot uses the Microsoft Prometheus model based on the foundation of OpenAI's GPT-4, a language model that has been fine-tuned through a combination of reinforcement learning methods. The conversational interface of the chatbot is similar to that of Chat GPT. It is capable of interacting in languages and dialect variations.

Google Gemini

Gemini is a Google-developed AI system that succeeds Bard that focuses on generating high-fidelity, diverse, and novel images using generative models. It employs generative adversarial networks (GANs) and advanced training techniques to create realistic and unique images. The primary goal of Gemini is to push the boundaries of image synthesis and enable applications in art, design, and entertainment.

Google's Bard, a chatbot like ChatGPT, was powered by Gemini Pro, which made it capable of more advanced reasoning and planning.

Bard and Gemini represent two distinct AI projects with unique goals and applications. While Bard streamlines the machine learning pipeline through reinforcement learning, Gemini excels in generating high-fidelity and diverse images using generative models. In 2024, Google unified Bard and Duet AI under the Gemini brand.

Amazon Q

Amazon Q is a ChatGPT-style chatbot designed for business users that is available as part of Amazon's market-dominating AWS cloud platform. It's aimed at people who use AWS at work, including coders, IT administrators, and business analysts. In response to typed requests, it will

81

CHAPTER 2   GENAI IN THE SPOTLIGHT

help developers write code, answer questions about how to use AWS cloud services for administrators, and generate business reports by tapping into QuickSight, a business intelligence platform in AWS.

Beyond this, Amazon Q has many other capabilities.

The new chatbot, from Amazon is now incorporated within Amazon Connect, a customer service platform that aids agents in resolving support inquiries effectively. Amazon Q is adaptable to company needs through accessing data sets and tailoring its functionalities for various staff members. It boasts security measures to ensure approval from IT managers.

Amazon Q incorporates intelligence models within its system, like Amazon's Titan language model and LLM models developed by the start-ups Cohere and Anthropic that are rivals of OpenAI.

Amazon Q has introduced the Amazon Q Apps to empower the development of applications powered by AI and natural language for easy sharing, among users.

It offers responses to queries and concerns and creates content to address issues effectively.

- Grasps the details about your company information and systems

- Tailors interactions to suit your position and access levels

- Designed for safety and confidentiality

When someone sends a question to the Q bot, the response can come from a model selected by the company and be automatically directed to the most suitable system.

Amazon Q aims to boost individual's efficiency and offers a range of features to help them in their tasks.

Filtering with metadata involves utilizing document characteristics to tailor and manage the chat interactions for end users, a feature exclusively when utilizing the Amazon Q API.

Fine-tuning for relevance involves leveraging document characteristics to enhance the generation of responses tailored to content within a clients platform.

Verifying responses by citing the source using Amazon Q attributions is essential for ensuring credibility and accuracy.

Allowing users to upload files in chat enables them to directly share files within the conversation and utilize the data for online activities.

Features sample quick prompts to enable end users of the capabilities of their Amazon Q web experience.

Creation of apps in natural language through conversations into reusable apps.

In this chapter, we introduced many aspects of generative AI as it has been reshaping industries and gained relevance since the pandemic. The ascension of conversational AI has unfolded an exciting new era of content-generative AI. In the next chapter, we will introduce many aspects of how generative AI has been envisioned in a new commerce, along with disruptions brought on a global scale by the pandemic and how they have impacted the software development lifecycle.

CHAPTER 3

Opportunities and Impacts of GenAI

The development of AI in the private sector has recently attracted much attention with people having different opinions on it, some seeing it as a revolution while the others are worried about the possible consequences. AI applications have been identified as powerful solutions in various sectors, transforming activities, improving performance, and presenting possibilities. These advancements are driving productivity, setting new standards of efficiency, and providing a competitive edge, thus making AI an essential part of today's organizations.

While artificial general intelligence is intended to develop computers that are able to learn and evolve similar to human beings, generative AI only aims to create new material with little or no contribution from humans. Despite the clear differences in their objectives, both subdomains of AI are motivated by the purpose of having machines be able to make more independent and innovative choices.

In fact, some scientists believe that creating generative AI is a critical prerequisite to developing AGI because any machine that is capable of creating its own content will be better able to meet human criteria for intelligence and learning.

CHAPTER 3   OPPORTUNITIES AND IMPACTS OF GENAI

The Impact of Generative AI on Modern Industries

Generative AI represents a cutting-edge frontier in the artificial intelligence domain, making significant strides with its ability to reshape numerous industries through its creative prowess. By leveraging machine learning algorithms, it can generate new content—whether images, sounds, or text—that mirrors the complexity and richness of original data. This technology paves the way for new forms of artistic and expressive innovation by digesting vast data sets and crafting entirely new, unique outputs. A quintessential example is its ability to create lifelike images of human faces that don't exist, drawing from its training on extensive facial imagery.

Furthermore, generative AI's influence extends across various creative fields such as visual arts, music, literature, and video production, proving itself a valuable asset for developers, artists, and creators alike. The rise of user-friendly platforms and tools has democratized access to this technology, empowering more people to harness its capabilities for their creative endeavors.

In tandem with generative AI, large language models (LLMs) have emerged as another powerhouse within AI, revolutionizing how machines comprehend and interact with human language. As a subset of natural language processing, LLMs utilize advanced machine learning and deep learning networks to analyze language patterns. Trained on colossal data sets comprising books, articles, and web content, these models strive to understand the intricacies of language structure. Through language modeling, LLMs can predict word sequences, paving the way for applications in machine translation, text classification, and sentiment analysis. The synergy between generative AI, AGI, and LLMs holds the promise of spawning potent and transformative tools across a multitude of sectors.

The pandemic caused by COVID-19 impacted many sectors of society on a global scale related to shipment delays, capacity issues, and increased consumer demand during the crisis.

The use of digital technology-based solutions was truly important during the pandemic. These included the Internet of things (IoT), next-generation 5G networks, and AI that uses deep learning, big data analytics, blockchain, and robotic technology, all of which has resulted in an unprecedented opportunity for the progress of telemedicine at scale.

AI was used extensively during the pandemic to improve COVID-19 detection and diagnosis. Some artificial intelligence algorithms have been created and utilized as an early screening tool for suspected instances, allowing individuals at high risk to be isolated or verified with laboratory-based tests. Furthermore, an AI-based triage system in the form of an online medical "chatbot" was deployed in the context of this epidemic to assist individuals in recognizing early symptoms and the need for hand cleanliness, which potentially lowered the burden of clinicians.

In the area of medicine and healthcare, AI performed admirably in diagnosis, prognosis evaluation, epidemic prediction, and drug research for COVID-19. AI had the ability to greatly improve existing medical and healthcare system efficiency during the COVID-19 pandemic.

The following sections enumerate some instances of AI usage in approaches taken during the COVID-19 pandemic, particularly for diagnosis, epidemic trend estimates, prognosis, and the development of safe and effective medications and vaccines.

Chest CT Images

The use of convolutional neural networks (CNNs) and deep learning has been on the rise because of their efficiency in image recognition especially. Almost all the studies (n = 18) conducted in this review developed AI models based on a CNN that was able to identify COVID-19 from non-COVID pneumonia with a high degree of accuracy using chest CT images for detection. It showed an accuracy of 70% to 99.87%, a sensitivity of 73% to 100%, a specificity of 25% to 100%, and an AUC of 0.732 to 1.000.

As described in the article "Artificial Intelligence for COVID-19: A Systematic Review," published in September 2021, Mei et al. created a combined CNN model quickly detects COVID-19 in patients by integrating chest CT results with clinical symptoms, exposure history, and laboratory testing. Furthermore, Mishra et al. suggested a decision-fusion strategy that blended each deep CNN model's predictions and produced outcomes of more than 86% for all performance measures considered. Three studies discovered that AI models had greater test accuracy, sensitivity, and specificity than radiologists and that with the use of AI, radiologists produced diagnoses at much quicker rates with better diagnostic performance.

Chest X-Rays

Although CT images are very efficient in revealing COVID-19, costs for the images and the radiation exposure are relatively high. On the other hand, a chest X-ray is an inexpensive and efficient modality that can be used in the first evaluation of the suspected cases of COVID-19 infection, thus supporting the appropriate application of quarantine measures in the confirmed patients. Some research works have proposed different AI algorithms to identify and isolate features from chest X-rays for diagnosis of COVID-19 with high accuracy (71.90 to 99.92%), sensitivity (75 to 99.44%), specificity (71.80 to 100.00%), and AUC (0.81 to 0.999).

Predicting the Prognosis of COVID-19

The capacity to assess a patient's risk of deterioration during hospitalization was crucial for allocating medical resources efficiently and ensuring adequate care for patients during the COVID-19 pandemic. We discovered various AI models based on chest CT scans that properly measured COVID-19–related lung anomalies and assessed disease severity and prognosis. Some research found that deep learning models could predict the likelihood of COVID-19 patients having critical illness based on clinical, laboratory, and radiological parameters at the time of hospital admission.

As cited in the article "Artificial Intelligence for COVID-19: A Systematic Review," published in September 2021, Iwendi et al. created a model that predicts the severity and potential consequences of COVID-19 cases based on geographical, travel, health, and demographic data from patients. In general, AI models predicted critical cases of COVID-19 with an accuracy of 74.4 to 95.2%, a sensitivity of 72.8 to 98.0%, a specificity of 55 to 96.87%, and an AUC of 0.66 to 0.997 (Table 3-1). Early diagnosis and treatment of COVID-19 patients may improve their prognosis and minimize death.

Table 3-1. Application of AI in Predicting COVID-19 Prognosis

References	Algorithm	Subjects	Objective	Results
Assaf et al. (68)	ANN, Random Forest and CRT	389 COVID-19 patients	Predict severity of COVID-19	Accuracy, 92%; Sensitivity, 88%; Specificity, 92.7%
Li et al. (70)	POI and iHU	196 COVID-19 patients	Predict severity of COVID-19	Sensitivity, 93.67%; Specificity, 88.05%; AUC, 0.97
Liang et al. (71)	Deep learning survival cox model	1,590 COVID-19 patients	Predict severity of COVID-19	Concordance index 0.894; AUC, 0.911
Yao et al. (75)	SVM	137 COVID-19 patients	Predict severity of COVID-19	Accuracy, 81.48%
Yu et al. (76)	DenseNet-201,SVM model	202 COVID-19 patients	Predict severity of COVID-19	Accuracy, 95.2%; Sensitivity, 91.87%; Specificity, 96.87%; AUC, 0.99
Iwendi et al. (69)	Random forest	–	Predict severity of COVID-19	Accuracy, 94%; Sensitivity, 75%; F1 score 86%
Abdulaal et al. (12)	ANN	398 COVID-19 patients	Predict mortality risk of COVID-19	Accuracy, 86.25%; Sensitivity, 87.50%; Specificity, 85.94%; AUC, 0.9012
Ma et al. (72)	Random Forest and XGboost	292 COVID-19 patients	Predict mortality risk of COVID-19	AUC 0.9521
Mushtaq et al. (73)	CNN	697 COVID-19 patients	Predict severity and mortality risk for COVID-19	For mortality, the AUCs were 0.66, for critical COVID-19, the AUCs were 0.77
Wu et al. (77)	LASSO logistic regression model	110 COVID-19 patients	Predict mortality risk of COVID-19	Sensitivity, 98%; Specificity, 91%; AUC, 0.997
Cheng et al. (78)	Random Forest	1,987 COVID-19 patients	Identify patients at risk of ICU transfer within 24 h	Accuracy, 76.2%; Sensitivity, 72.8%; Specificity, 76.3%; AUC, 0.799
Fu et al. (79)	LASSO, mRMR, SVM	64 COVID-19 patients	Identify the progression of COVID-19	Sensitivity, 80.95%; Specificity, 74.42%; AUC, 0.833
Wu et al. (74)	ADASYN, Logistic Regression	426 COVID-19 patients	Predict severity risk for COVID-19	Accuracy, 74.4–87.5%; Sensitivity, 75–96.9%; Specificity, 55–88%; AUC, 0.84–0.93
Xiao et al. (80)	MIL, ResNet34	408 COVID-19 patients	Predict severity risk for COVID-19	Accuracy, 81.9%; AUC, 0.892

ANN, Artificial Neural Network; RF, Random Forest; CRT, Classification and Regression Decision Tree; POI, portion of Infection; IHU, average infection Hounsfield unit; SVM, Support Vector Machine; LASSO, Least Absolute Shrinkage and Selection Operator; mRMR, Minimum Redundancy Maximum Correlation; ADASYN, Adaptive Synthetic Sampling; MIL, Multiple Instance Learning.

CHAPTER 3   OPPORTUNITIES AND IMPACTS OF GENAI

Predicting the Epidemic Trends of COVID-19

COVID-19 was declared a pandemic by the World Health Organization (WHO) in March 2020. As the COVID-19 pandemic progresses, it is critical to focus on developing prediction models to assist policymakers and health managers in allocating healthcare resources and preventing or mitigating outbreaks. We discovered nine papers that attempted to forecast the COVID-19 pandemic trend (Table 3-2).

Table 3-2. Application of AI in Predicting the Epidemic Trends of COVID-19

References	Algorithm	Country	Objective	Results
Alsayed et al. (81)	GA, SEIR, ANFIS	Malaysia	Estimate the infection rate, epidemic peak, and the number of infected cases	Infection rate is 0.228 ± 0.013, NRMSE 0.041, MAPE 2.45%, R2 of 0.9964
Ayyoubzadeh et al. (82)	LSTM, linear regression	Iran	Predict the incidence	RMSE: LSTM, 27.187 (SD 20.705); Linear regression, 7.562 (SD 6.492)
Mollalo et al. (83)	MLP neural network	The US	Predict incidence rates	RMSE, 0.722409; MAE. 0.355843; correlation coefficient 0.645481
Shahid et al. (84)	ARIMA, SVR, LSTM, Bi-LSTM	Ten countries	Predict confirmed cases, deaths, and recoveries	BI-LSTM generates lowest MAE and RMSE values of 0.0070 and 0.0077 in China; r2_score 0.9997
Zheng et al. (85)	ISI, NLP, LSTM	China	Analyze the transmission laws and development trend	Obtain MAPEs with 0.52, 0.38, 0.05, and 0.86% for the next 6 days in Wuhan, Beijing, Shanghai, and countrywide, respectively
Arora et al. (86)	LSTM	India	Predict daily and weekly positive cases	Daily predictions MAPE <3% and weekly predictions MAPE <8%
Chimmula and Zhang (87)	LSTM	Canada	Predict the trends and possible stopping time of COVID-19	For short term predictions, RMSE, 4.83; accuracy, 93.4%. For long term predictions, RMSE, 45.70; accuracy, 92.67%
Ribeiro et al. (88)	SVR, stacking-ensemble learning, ARIMA, CUBIST, RIDGE, and RF	Brazil	Forecast the cumulative confirmed cases	sMAPE in a range of 0.87–3.51, 1.02–5.63, and 0.95–6.90% in 1, 3, and 6-days-ahead, respectively
Shastri et al. (89)	Variants of LSTM	India, The USA	Forecast the confirmed cases and death cases	Achieved accuracies of 97.82, 98, 96.66, and 97.50%, MAPE of 2.17, 2.00, 3.33, 2.50 for India confirmed cases, USA confirmed cases, India death cases and USA death cases, respectively

GA, Genetic Algorithm; SEIR, Susceptible-Exposed-Infectious-Recovered; ANFIS, Adaptive Neuro-Fuzzy Inference System; LSTM, Linear regression and long short-term memory; R_0, Reproductive number; MLP, Multilayer perceptron; RMSE, Root-mean-square error; NRMSE, Normalized root mean square error; MAE, Mean absolute error; ISI, Improved susceptible-infected; PSO, Particle Swarm Optimization; MAPE, Mean Absolute Percentage Error; NLP, Natural Language Processing; ARIMA, Autoregressive Integrated Moving Average; CUBIST, Cubist Regression; RF, Random Forest; RIDGE, Ridge Regression; SVR, Support Vector Regression.

Six of these investigations employed long short-term memory (LSTM) models, with or without other models, to estimate COVID-19 incidence, confirmed cases, fatalities, recoveries, development trends, and potential stop time. Alsayed et al. employed the Susceptible-Exposed-Infectious-Recovered (SEIR) model in conjunction with machine learning to forecast the progression of the epidemic and estimate the number of unreported infections. Mollalo et al. investigated the use of multilayer perceptron (MLP) artificial neural networks in replicating cumulative COVID-19 incidence at the county level across the continental United States. Shahid et al. (84) proposed prediction models such as support vector regression (SVR), autoregressive integrated moving average (ARIMA), LSTM, and bidirectional long short-term memory (Bi-LSTM) to predict confirmed cases, deaths, and recoveries in 10 major countries affected by COVID-19.

Zheng et al. suggested an improved susceptible-infected (ISI) model for estimating infection rates, analyzing transmission laws, and predicting development trends. Ribeiro et al. employed a variety of machine learning models to anticipate the cumulative confirmed cases of COVID-19 in 10 Brazilian states with a high daily incidence, ranking the models based on accuracy. These findings might have a wide-ranging impact on COVID-19 control and prevention efforts.

Drug Discovery and Vaccine Development for COVID-19

With the rise of COVID-19 and few proven effective treatments available, it was critical to develop antiviral medicines and vaccines against SARS-CoV-2. Traditional approaches often take a long time to generate a treatment or vaccination; however, AI techniques were used to find promising pharmaceuticals and develop effective and safe COVID-19 vaccines.

Drug Repurposing

Drug repurposing is the practice of applying established medications to new therapeutic indications, which has been shown to be an effective drug development technique for lowering development costs and simplifying drug approval procedures. AI algorithms were trained and then utilized to screen current medications for efficacy in the treatment of COVID-19.

Drug Development

Zhang et al. developed a 3D protein model of 3CLpro, utilized a deep learning approach to discover protein-ligand interactions, and then gave probable chemical and tripeptide lists for 3CLpro. Batra et al. used machine learning and high-fidelity ensemble docking to discover 75 FDA-approved ligands and 100 additional compounds from medication data sets as possible COVID-19 therapeutics. Joshi et al. employed deep-learning algorithms to screen natural compounds and discovered that two compounds, Palmatine and Sauchinone, formed a very stable combination with Mpro, which might be exploited for medicinal development against SARS-CoV-2. Ton et al. employed Deep Docking (DD) to screen 1.3 billion molecules from the ZINC15 library and find the top 1,000 candidate ligands for the SARS-CoV-2 Mpro protein.

Vaccine Development

Without available effective medical therapy, developing an effective and safe vaccine is a crucial strategy for dealing with the highly contagious disease caused by the SARS-CoV-2 coronavirus. By analyzing the whole SARS-CoV-2 proteome, Ong et al. used a machine learning algorithm to predict which S proteins, nsp3, 3CL-pro, and nsp8-10, were critical to viral adherence and host invasion. The SARS-CoV-2 S protein has the greatest protective antigenicity score and was designated as the most promising vaccine candidate; also, the nsp3 protein was chosen for future study. The anticipated vaccine targets have the potential to lead to the development of a COVID-19 vaccine; nevertheless, they must be further examined in clinical research.

Enterprise Modernizing and Faster Coding

We are witnessing a massive disruptive movement akin to the birth of Alan Turing's Enigma, Margaret Hamilton's software engineering, or the first website published in a new world now known as the Internet.

CHAPTER 3   OPPORTUNITIES AND IMPACTS OF GENAI

After much time and work over the last decade, we are beginning to see a broad future in which concepts utilized in artificial intelligence for many years will become available to a huge portion of the population.

Today, we are continuously utilizing artificial intelligence, either directly or indirectly. Considering direct use of AI, I can't remember a time when I didn't interact with a recommendation generated by my user profile, searches based on my previous week's interests, or chatbots that have the ability to facilitate my day and provide me with access to services that were available only during business hours. I'm not sure when was the last time I had to wait in long lines and reschedule appointments.

AI is present in all indirect services, even for the most analog person imaginable. It is used by mobile services to find the nearest tower with a stable signal, television channels to filter their content, and the second level of authentication for applications. It can also appear in the text of a print newspaper, an image of an advertising campaign at a bus stop, or public transportation schedules when they are updated.

The coronavirus pandemic has brought AI to the forefront, as it has been introduced in vaccine trials, as well as computer vision technologies for reading temperature in large cities or counting the use of protective masks.

We have made progress in the software development life cycle (SDLC) through technological immersion in our environment, following a long period of science, studies, and errors. Going forward, we will modify the way we develop, maintain, and create digital platforms.

What was once so far away has recently gone viral through ChatGPT, shedding light on a collection of LLMs and GenAI, both of which are rapidly becoming a part of our daily lives.

Generative AI is changing our understanding of how AI is used. It represents the democratization of powerful resources for a diverse group of nontechnical users. It also makes tasks easier for professionals by reducing application development time and improving quality.

CHAPTER 3   OPPORTUNITIES AND IMPACTS OF GENAI

A new class of team-integrated toolkits will feature platforms that include Amazon Q Developer,[1] ChatGPT, GitHub, Copilot, Midjourney, and Tabnine. They will represent the next generation of SDLC.

There are many business teams with enormous backlogs who need to accelerate their digital transformation by leveraging GenAI capabilities. Amazon Q Developer has revolutionized software development significantly. One example is the transition to Java 17, where the upgrade time can be reduced from 50 days to hours, resulting in saving an estimated 4,500 developer-years and generating $260 million in efficiency gains. These numbers are crazy but real.

In this chapter, we will examine the impact of AI on how we work, collaborate, create software, and plan for the future of our teams in this new era. Prepare to discover how generative AI is speeding, smartening, and improving the software development business.

A New Hybrid World

We've arrived in a new era: the hybrid world. The term *convergence* has gained a lot of attention in the development world over the last decade, with a focus on information and operations technology teams and the opportunities that come with them.

We have recently observed an increase in contact between the physical and virtual worlds as a result of Industry 4.0 and its Digital Twin and Cyber-Physical System initiatives. The Industrial Metaverse, also known as Digital Twin 2.0, has recently given the concept new life.

In the preceding scenarios, we can see that the first one is impossible due to a technology gap and a lack of mechanisms for integrating human expertise in two different fields. In the second, we encountered a situation where the technology needed to support the complete scope was not mature enough.

---

[1] As of April 30, 2024, Amazon CodeWhisperer is part of Amazon Q Developer, including inline code suggestions and Amazon Q Developer security scans.

CHAPTER 3   OPPORTUNITIES AND IMPACTS OF GENAI

The storm produced by ChatGPT has been significant because it uses an interface that people have been using and interacting with for decades. Silicon computers are noted for their early steps using prompts and SMS, the same as today's conversational interfaces like WhatsApp.

Movies and series such as *A.I. (Artificial Intelligence), 2001: A Space Odyssey*, and *Better Than Us* have often portrayed AI as a formidable and independent agent that eventually rebels against humanity. This portrayal has fostered a widespread perception that AI exists as an adversary to human beings, especially in terms of job opportunities, rather than being a capable partner that can contribute to mutual success.

Since the start of AI as a field of research, the concept of humans and AI working together has been a driving force. Early AI visionaries envisioned a collaborative partnership in which AI would improve human cognitive capacities to tackle complicated challenges, make better judgments, and help people achieve their goals more successfully.

So, what exactly is the new hybrid world made up of?

Today, the digital transformation market is accelerating significantly. The global economy now breathes and sweats electronically. To keep up with business teams that frequently have large, sometimes unclear, and undefined backlogs, businesses need to recognize the importance of technology as a tool.

We now rely on generative artificial intelligence to speed up the numerous points of contact between humans and creativity. This technology breathes, dreams, and sweats human essence to make our days easier. We can now access tools that help with creative tasks like unlocking critical moments or dealing with large and repetitive tasks on a regular basis.

Meanwhile, we believe that by combining the power of AI and human intellect, we will create a new professional whose product will be hybrid in nature. AI will speed the beginning of the creation process, making it significant from the very start of production.

In the following topic, we will explore how rapid development will affect business teams and the future potential of this movement.

CHAPTER 3   OPPORTUNITIES AND IMPACTS OF GENAI

Getting Even Faster

After paving the road for the future of hybrid teams, we must now understand how speed will affect a software developer's process and the ecosystem that surrounds them.

When a team first implements the GenAI culture, there will be a considerable cultural impact on the organization's structural units. This extends beyond development and creative teams to the business team.

We believe that if teams have a robust tool in their toolkits, it will make their project specification round more complete. It will allow for speedier narrative writing and the development of more scenarios to test novel situations, resulting in broader product coverage.

With this adjustment, discovery cycles could alter drastically, with more rounds, shorter execution durations between them, and a longer period for digesting the material.

In terms of behavior, increased speed will influence the team's morale. Following the lean startup methodology allows developers to make mistakes and fix them quickly, resulting in a more creative team with less fear of failure.

By fixing issues promptly, the cost and time required to present new scenarios for collective decision-making will be reduced. On the other side of the balance, there will be more solid inputs and richer backlogs, with less distortion in the business vision, because the tools may be asked to translate in several ways until the right spectrum is reached.

This means that with more speed and a lower cost, there's a rich R&D laboratory that does not follow the classic models of GenAI, but rather a playground style in which vast elements may be altered quickly.

With a small team, proof of concepts (POCs) can be produced in days, leading to minimal viable products (MVPs) within a business week.

This acceleration in the product development process may enable businesses to respond faster to market and customer demands. Consider the following scenario: Principal complaints are sent via a production pipeline in order of criticality. The team can determine whether the pipeline is consistent, and all documentation can be prefilled and ready for the development or creative team to make changes on the fly.

In terms of machine learning, the viability of various models can be determined using simulated PoCs in a GenAI environment. Using an advanced LM tool, the building of a specific regression model can be requested to solve a certain problem. This method provides a first, unrefined deliverable that can be evaluated to determine the authenticity and effectiveness of the suggested solution. This method will allow evaluation of the potential of various ML models while also indicating areas that may require additional modification or optimization for best results.

Still, a chapter might be written on ML about how this field of study will stand to gain mostly from two scenarios: doing ML model training with LLMs and producing data sets using synthetic data approaches.

The Rise of Cyborgs

Increased speed will undoubtedly have a substantial impact on the software development production chain. This transition will have a direct impact on project management cycles, as well as the efficiency and performance of business teams. As we gradually integrate GenAI platforms into our teams, our view on value and time will shift, resulting in a new way of conducting software development.

In this section, we will look at the specific personas who will serve as agents of this transition. We will investigate their particular skills and capacities, as well as the far-reaching implications of incorporating these personas into the future of software development.

Understanding their responsibilities and contributions will allow us to better predict and negotiate the changes brought about by this exciting new era of technology.

Cyborgs in a Nutshell

The term *cyborg* has been widely used to designate a variety of notions and ideas concerning the union of humans and technology. The word was first used in the 1960s to describe humans who had been changed with electronic or biomechanical components to improve their physical and cognitive capacities. However in current times, the concept of a cyborg has expanded to include the integration of humans with digital parts that function as an extension of their body.

In the context of the current subject matter, a contemporary cyborg is a professional who uses GenAI to assist with their daily tasks, resulting in a synergy between human and artificial intelligence. This combination enables the cyborg to benefit from the skills and competences of both, leading to better performance and efficiency at work. Simultaneously, the ongoing interaction between humans and AI promotes their growth and evolution, resulting in a circle of mutual learning and adaptation.

Specialized Cyborg

Modern cyborgs can be found in a variety of disciplines and industries, including software development, product management, and business. These experts are distinguished by their versatility, ongoing learning capacity, and analytical and creative abilities.

Furthermore, cyborgs emphasize collaboration and communication, both within teams and with AI. This synergy enables cyborgs to address challenges and discover novel solutions while benefiting from the advantages of generative artificial intelligence.

In the following sections, I'll provide thorough explanations of the personas of each of these cyborgs, as discussed with ChatGPT using the GPT-4 model. I asked it *what the three most important attributes a colleague might have of each sort of cyborg*. To contextualize the debates, I typed the text up to this point directly into the prompt as input.

Business Cyborg

The *business cyborg* would be a strategy and management expert who uses AI to make informed decisions and promote business growth. With extensive market and industry expertise, this cyborg could detect emerging possibilities and address challenges in novel ways. The business cyborg is adaptable and continually learning, and it employs GenAI to evaluate trends, predict changes, and strategically position itself.

According to ChatGPT, it must feature the following skills:

> **Strategic vision**: A business cyborg must be able to analyze the big picture, identify opportunities and challenges, and devise strategies using GenAI as support.
>
> **Knowledge of the market and industry**: A business cyborg must have a deep understanding of the market and industry in which it operates, using GenAI to identify trends and emerging opportunities.
>
> **Interpersonal and leadership skills**: A business cyborg must be able to communicate effectively and inspire and motivate their team, establishing a synergetic relationship between humans and GenAI.

Development Cyborg

The *development cyborg* would be a highly skilled IT professional who could combine their technical and human skills with the power of GenAI. This cyborg could assess and solve complex problems, as well as design novel solutions, thanks to their extensive understanding of programming languages, frameworks, and tools. They also will have interpersonal abilities, such as communication and cooperation, which allow them to engage with colleagues while maximizing their cognitive strength.

According to ChatGPT, this type must feature the following skills:

> **Problem-solving ability**: A development cyborg must have skills to analyze and solve complex problems, using GenAI as support.
>
> **Technological adaptability**: A development cyborg must always be up-to-date with the latest trends and technologies, adapting to new tools and approaches.
>
> **Teamwork and communication**: These professionals must know how to work in a team and communicate effectively, both with human colleagues and with their new colleague, GenAI.

Product Cyborg

The *product cyborg* would be a professional who combines human skills like empathy, creativity, and communication with GenAI's analytical and data processing capabilities. This persona could be identified by the creative staff. The integration enables the company to create novel, user-centered solutions, constantly upgrading products and keeping them relevant in today's competitive market.

According to ChatGPT, it must feature the following skills:

> **Project management and prioritization skills**: These professionals need to coordinate projects efficiently, establish priorities among tasks and resources, and leverage GenAI to optimize process and identity improvement points throughout the product lifecycle.

Communication and collaborative work:
A product cyborg must know how to communicate clearly and efficiently with development and business teams, as well as GenAI. In this way, it is possible to ensure synergy between all parties involved and promote a collaborative work environment.

Analytical approach and data-driven orientation:
A product cyborg must be able to analyze data and metrics to make informed decisions about product development and improvement. Using GenAI, the professional can identify trends, patterns, and valuable insights to optimize performance and user satisfaction with the product.

The Secret Sauce

Cyborg-driven hybrid work has a clear goal: to use AI to turn novel processes into efficiency.

Efficiency will result not just from speedier development but also from the creation of more solid digital platforms. By deploying cyborgs to construct digital projects, we will gain more cognitive strength from the interplay of humans and AI.

We can perceive this picture from different angles. On one side, we have a human conductor who is less burdened and brighter than before. On the other, we have a quick and tireless engine with abilities in a variety of circumstances.

It must be emphasized that if teams do not integrate GenAI into their SDLC, the train will derail. Those who opt to compromise quality for speed may end up with fragile goods that lack effective lifecycle tracking.

The methods mentioned thus far will make applications more robust by allowing more human cognition to be applied to extend linkages and analytical power over situations.

When it comes to hard skills, there can be more concentration on testing, which includes security, usability, and user journeys. Several articles can be generated that will cover most use cases, as well as meet customer needs that have been mapped using AI.

Using a digital brain for pair programming will result in more detailed documentation, making future maintenance of new apps easier. The time it takes to detect and resolve issues in production environments will be shortened.

The software industry is predicated on transformation, so we must invest in our teams' basic education. To build people with hybrid qualities, we must believe in professionals' soft skills and support their inventiveness and adaptability. That will be the secret sauce.

Prompt Engineering

With the explosion of a vast number of LLMs, prompt engineering has received lot of attention as a new and relevant area of knowledge in experience with generative AI solutions.

Prompt engineering is the practice of guiding GenAI solutions to produce the desired results. Even though generative AI aims to emulate humans, it requires precise instructions to produce relevant, high-quality results. Prompt engineering involves selecting the most relevant formats, phrases, words, and symbols to let AI connect with users more meaningfully. Prompt engineers utilize creativity and trial and error to develop a collection of input texts, ensuring that an application's generative AI performs as intended.

What Is a Prompt?

A *prompt* is a text entered in a natural language format that guides the generative AI to complete a certain task. Generative AI is an artificial intelligence technology that generates new content, such as conversations, stories, images, music, and videos.

It is powered by very large ML models that employ deep neural networks pretrained on massive amounts of data.

LLMs are highly adaptable and can perform a variety of jobs. For example, they can summarize papers, finish phrases, answer queries, and translate languages well. For specific user input, the models anticipate the optimal result based on prior training.

However, since they are so open-ended, consumers may engage with generative AI solutions using a variety of input data combinations. The AI language models are quite strong and do not require much to begin producing content. Even one word is enough for the algorithm to provide a thorough answer.

Not all types of input produce useful results. Generative AI systems rely on context and extensive information to provide correct and appropriate replies. Prompts generated methodically obtain more relevant and useful results. So, prompt engineering is basically the process of repeatedly refining prompts until the AI system produces the results you want.

What Makes Prompt Engineering So Relevant?

Since the debut of generative AI, there has been a considerable surge in prompt engineering positions. Prompt engineers bridge the gap between end users and a large language model. They identify scripts and templates that users can alter and fill out to achieve the greatest results from the language models. These engineers experimented with several forms of inputs to create a prompt library that application developers can utilize in a variety of situations.

Prompt engineering increases the efficiency and effectiveness of AI applications. Open-ended user input is often encapsulated within a prompt before being passed to the AI model.

Consider AI chatbots. A user may submit an incomplete problem statement, such as "Where to rent a car?" Internally, the application's code includes a designated prompt that reads, "You are a sales assistant for a sales rental company. A user from Sao Paulo, Brazil, is asking you where to rent a car. Please respond with the three nearest rental car locations that currently offer the requested car." The chatbot then provides more relevant and accurate information.

Here are some of the advantages of the prompt engineering approach:

> **Greater developer control**: Prompt engineering allows developers to have more control over how users engage with AI. Appropriate prompts give intent and context for large language models. They assist the AI in refining the output and presenting it succinctly in the appropriate format.
>
> They also prevent users from abusing the AI or demanding something it does not understand or cannot manage correctly. For example, in a commercial AI application, a company may want to prevent users from creating inappropriate material.
>
> **Improved user experience**: Users may eliminate trial and error while still receiving clear, accurate, and relevant results from AI technologies. Prompt engineering enables users to get appropriate results in the first prompt. It helps to reduce bias that may be contained in the training data of large language models due to existing human bias.

Also, it enhances the user's interaction with AI so that the AI can understand the user's request with the least possible words. For instance, the requests to break down a legal file and a news item and get two different outputs, depending on the prompt style and the tone. This is the case even if both the users are simply telling the application "Summarize the following text," for instance.

Greater flexibility: Higher degrees of abstraction enhance AI models and enable enterprises to develop more adaptable tools at scale. A prompt engineer can design prompts with domain-neutral instructions that emphasize logical relationships and broad patterns. Organizations can readily utilize the prompts throughout the company to maximize their AI efforts.

For example, to identify possibilities for process improvement, the prompt engineer can design several prompts that train the AI model to detect inefficiencies using wide signals rather than context-specific data. The prompts may then be utilized across several work processes and business areas.

Which Use Cases Can Take Advantage?

Users may improve their experience when applying prompt engineering in large language models. Here are some possible examples:

Contextual expertise: Prompt engineering is critical in applications that demand AI to respond with subject matter expertise. A prompt engineer with relevant knowledge may lead the AI to the proper sources and formulate a better answer based on the inquiry.

For example, in the field of medicine, a doctor might provide a differential diagnosis for a challenging case using a prompt-engineered language model. A doctor essentially must enter the symptoms and patient information. The AI will use tailored prompts to identify the most likely diseases related to the entered symptoms. By using further patient information, a more relevant selection is narrowed down.

Critical thinking: This relies on a language model capable of supporting complicated challenges. To do so, the model examines data from many perspectives, assesses its reliability and makes informed conclusions. Prompt engineering improves a model's data analysis capability.

For example, in decision-making circumstances, you may ask a model to list all feasible possibilities, analyze them, and propose the best answer.

Creativity: This helps to generate new ideas, thoughts, and alternatives. Prompt engineering may be used to improve a model's creative ability in a variety of situations.

For example, a writer may utilize a prompt-engineered model to assist in developing story ideas. The writer may ask the model to brainstorm probable characters, places, and narrative ideas before developing a tale around those features. Alternatively, a graphic designer may ask the model to provide a list of color choices that produce a specific feeling and then construct a design utilizing that range.

CHAPTER 3   OPPORTUNITIES AND IMPACTS OF GENAI

What Are Applied Techniques?

Prompt engineering is an ever-changing process. To be a good prompt engineer one has to possess language skills and the ability to come up with new ideas to craft the right prompts for the generative AI tools and get the expected output from them.

Here are some more examples of techniques that prompt engineers use to improve their AI models' natural language processing (NLP) tasks:

> **Chain-of-thought prompting:** This approach for restructuring a big question into a number of reasonable steps can look like the steps of a thought process. This enables the model to solve problems in number a of stages instead of solving the problem all at once. This enhances its problem-solving skills.
>
> A number of chain-of-thought rollouts may be conducted in the course of a task especially where the task is complex and one of the rollouts can be chosen based on the closeness of the outcome to the actual outcome. If the rollouts are completely different, someone can help in organizing the thought process.
>
> For example, if the question is "What is the capital of EUA?" the model may perform several rollouts, and its responses may include "Washington," "The capital of EUA is Washington," and "Washington is the capital of EUA." Since all the rollouts yield the same answer as the input, Washington will be chosen as the output.

Tree-of-thought prompting: This approach is a special case of chain-of-thought prompting. It leads the model to come up with one or more possible next actions. It is different from conventional the chain of thought in the sense that it operates in a tree search manner where the model through is each run of the possible next steps.

For instance, take the question, "What are the effects of climate change?" The model may first come up with the following potential future stages, for instance, "List the environmental effects" and "List the social effects. " It would then describe more about each of them in the subsequent stages.

Maieutic prompting: This is similar to the tree-of-thought prompting. The model is provided with a question that it has to answer and an explanation. It is then requested to elaborate on certain parts of the explanation. Those trees that offer wrong explanations are removed, or if they are right, then they are pruned. This enhances the performance of the model in challenging commonsense thinking tasks. For instance, if the question is "What are clouds made of?" the model may answer, "Clouds are made up of tiny water droplets, ice crystals, or a combination of both, depending on temperature and atmospheric conditions." It may then go on to discuss various aspects of this detailing more about water droplets and ice crystals definitions.

Complexity-based prompting: This is a prompt engineering approach that comprises coming up with many chains of thought. It chooses the rollouts with longest number of steps and then the most often repeated end. For example, if the query is a difficult math question, the model may perform many rollouts, and each rollout may contain several calculation steps. It would also consider the rollouts with the longest chain of thought, and in this case, it is the most calculated actions. The rollouts that converge to a similar answer as other rollouts will be selected as the solution.

Generated knowledge prompting: This method makes the model generate the appropriate facts before continuing with the request. Subsequently, it proceeds to complete the request. Thus, since the model is conditioned on key facts, completion quality is often enhanced.

For instance, if a user requests the model to develop an essay on the impacts of releasing carbon dioxide, some of the outputs that the model may generate at the initial level may include the statements such as "releasing carbon dioxide causes climate change" and "releasing carbon dioxide results to loss of biodiversity." Then, it would develop on the topics of the essay.

Least-to-most prompting: In this case, the model is first asked to identify the individual components of a problem before working through the problem in its entirety. This technique makes sure that solutions to later subproblems can rely on the solutions to earlier subproblems.

For instance, let's take a user asking the model a math question like "Solve for x in equation 2x + 8 = 40. " The model might first identify the subtasks as "Subtract eight from each side" and "Divide by two. " It would then solve them and arrive at the final answer.

Self-refine to prompting: With this approach, the response provided is with revised a keeping request, in in it view has the request, response, and evaluation made by the model. The problem-solving process is iterated until such a point that is stated to be end the of the process. For instance, it can be for lack of tokens or time, or the model could output a stop token. For example, if a user asks a model to "Write a short essay on quantum chemistry," the model may produce an article, analyze it for lack of examples, and then rework it using related examples. This procedure would continue until the essay is considered satisfactory or a stopping requirement is fulfilled.

Prompt engineering: This approach involves structuring the prompt in a tip or a cue like, for example, desired keywords.

For instance, if the prompt is on the topic of medieval story, the prompt engineer can define some prompts in the form of certain words like "knight," "sword," and "shield. " Therefore, when the model is requested to compose a poem, it will include the specified terms that would make the model generate a poem that incorporates all these keywords.

What About Best Practices?

Good prompt engineering leads to providing instructions with context, scope, and expected reaction. Next, we'll go over some best practices:

Unambiguous prompts: To avoid misunderstanding by the AI, the user should provide unambiguous instructions that clearly explain the expected answer. For example, if you're asking for a novel synopsis, make it obvious that you want a summary rather than a full analysis. This enables the AI to focus solely on a request and produce an answer that is consistent with the goal.

Adequate context within the prompt: Users should include output requirements in their prompt input while keeping it to a specified format. Consider a user seeking a table with a list of the most popular Seattle grunge bands of the 1990s. To get the precise result, they should specify how many movies they want to be included and request table formatting.

Balance simplicity and complexity: Prompts should be kept balanced to avoid ambiguous, irrelevant, or unexpected responses. A short prompt may lack context, but a complicated prompt may mislead the AI. This is especially crucial for difficult topics or domain-specific languages that the AI may not be familiar with. Instead, using basic language and minimizing the size of the prompt will make the request easier.

Experiment and enhancement: Prompt engineering is an iterative process. It is critical to experiment with various concepts and test AI prompts to observe the results. It may take many attempts to optimize for accuracy and relevancy. Continuous testing and iteration of lower prompt sizes help the model produce better results. There are no set standards for how the AI generates information; therefore, versatility and adaptability are required.

Limitations and Concerns

Every time we have heard that new technology would replace human tasks, we've assumed this development would free up time for humans to ponder or conduct more complex work. Given the standard SDLC's implementation, testing, and deployment phases, the key requirement for a developer is knowledge of the programming language and the ability to code in it.

What tasks will humans perform now that GenAI models are writing program code? Is AI changing the way we work?

With all of these generative AI and LLM programs accessible to speed up software development, developers will need to learn a new skill: how to speak with AI.

GenAI models can already be copilots in a variety of languages—all you have to do is ask. No one needs to be fluent in a specific language to use AI during the SDLC process, but they must understand how it works and how to ask the proper questions to get an accurate response. As a result, the new core skill for a developer will be to master the art of writing successful AI instructions.

CHAPTER 3   OPPORTUNITIES AND IMPACTS OF GENAI

Generative AI and LLMs might be formidable technologies that will disrupt the IT sector, but many components of the SDLC require human intervention. While AI and automation can assist with specialized activities, some tasks, such as decision-making, require human intervention.

When it comes to software development, decisions are essential. Developers must be able to choose which features to create, how to build them, and when to release them. This necessitates a thorough understanding of both the problem at hand and the users for whom the product is intended.

While AI can help with some aspects of the process, it is not yet capable of completely replacing humans. This means that human developers continue to play an important role in software development.

The purpose of this section is to prompt you to consider some of the constraints and difficulties associated with employing GenAI models in the SDLC, as well as discuss how the code produced by AI code writers will comply with intellectual property rights.

The Asking Paradigm with GenAI Models

What makes generative AI and LLM popular in many fields is their capacity to interact using natural language processing (NLP), which means we can ask the AI questions in human languages rather than computer programming languages.

However, utilizing AI to assist developers in creating codes and obtaining what is desired can be a difficult undertaking because natural language can have multiple meanings for a single statement, making it difficult for AI models to understand what they're being asked to generate. Furthermore, despite the large amount of data used to train GenAI models, they may require a perfect match of the specific order to produces accurate responses.

CHAPTER 3 OPPORTUNITIES AND IMPACTS OF GENAI

Developers must grasp the art of writing efficient AI instructions and comprehend AI defects; otherwise, erroneous results may occur. Knowing the audience before a presentation allows the content to be tweaked so that the audience understands the intended messages.

This means that developers will require a thorough understanding of how GenAI models work to properly tell the AI to write code that matches their expectations.

Developers must already devote time thinking about and planning how to create logical prompts that aid the AI code writer in producing accurate code and ensuring that the created code adheres to industry best practices, standards, and guidelines.

As the AI-generated code grows in size and complexity, more human supervision is required to advise and assist the AI in making decisions on how to best apply changes.

Furthermore, understanding a computer programming language, including its strengths and limits, will be critical in providing AI with the instruction it requires to develop correct results and optimal code that uses the fewest resources possible.

Most IT learning platforms are overwhelmed with online content, courses, and tutorials that teach programmers how to write AI commands. These courses teach prompt engineering, which is viewed as a new career emerging as generative AI applications develop.

Startups and businesses are already hiring and training prompt engineers to use generative AI models in the software development process.

So far, it looks as if AI code generators will do well in Copilot work. However, human developers will remain as the main actors, as these models rely on their input and feedback to ensure that the created code matches the project's needs.

Legal Concerns

Companies that own digital products are accustomed to dealing with intellectual property and copyright issues, with legal advisors supporting them on each new software product launch, as well as on a daily basis.

Many GenAI models are trained utilizing data from publicly available code repositories, which may contain code from open-source licensed programs.

Would the usage of GenAI models to assist developers in building a portion or practically all of a program's source code constitutes a legal infringement of intellectual property (IP) or copyright rules? Or can you still protect your code with copyrights or patents?

It is highly likely that as you read this chapter, regulatory authorities are attempting to determine how to treat or even monitor property rights in source code created by GenAI models.

Let's analyze the GitHub Copilot, one of the most popular AI programming pairs, and what areas of intellectual property and copyright we should be concerned about.

Knowing Your Rights

According to GitHub, Copilot leverages OpenAI Codex to recommend code and complete functions in real time. Codex was trained on natural language text and source code from billions of publicly available sources, including code from GitHub's public repository.

Many public codes available on code repositories may incorporate open-source licensed code. An open-source licensed code is subject to limitations and conditions that may limit the company's ability to seek and enforce patent rights.

It should come as no surprise if an AI pair programming proposes an exact duplicate of the code used to train it in response to a prompted query. Many of them have admitted that the code generated by the tool occasionally references publicly available open source code on which it was trained. As a result, the license for the open-source code may apply to the code generated.

This condition may raise concerns and expose them to legal problems, such as copyright breaches. Of course, such a situation might be averted by reviewing the code with automated code-scanning tools or by implementing a code review process to detect and control the risk of committing infringements.

The large lawsuit against GenAI models for code creation is only the beginning, with many unsolved questions around intellectual property (IP) rights and ownership on the horizon. New licenses have been created to address some of these concerns. However, it may take some time for developers to embrace them and for GenAI models to be retrained with just licensed source code.

While GitHub's enterprise edition of Copilot may help with some IP issues, there are still many unsolved questions. The lack of traceability in GenAI tools complicates the identification of any open-source license violations.

As a result, it is recommended to utilize these tools with caution and manually scan or use scanning software on created code to detect any open source–licensed code and avoid accidentally violating open-source licenses and copyrights.

More Concerns

In an open letter signed by more than 3,100 people, including Apple co-founder Steve Wozniak and Elon Musk, tech leaders criticized San Francisco-based OpenAI Lab's recently announced GPT-4 algorithm, stating that development should be halted until we have a better understanding of the consequences of using GPT-4. Italy banned the use of OpenAI technology due to privacy concerns. We've already witnessed a hack of OpenAI, where chats and payment information were exposed.

As a result, there are numerous worries about the use of AI, which can be related to the SDLC we previously discussed. AI can provide various benefits, but it also poses certain risks that must be addressed. Some of the significant problems involving AI are listed in this section.

Ethical Implications

AI may bring ethical problems, particularly around privacy, security, and transparency. For example, AI-powered software may collect and handle massive volumes of data, possibly jeopardizing user privacy. Strong ethical rules are required to control the use of AI in the SDLC and ensure that the ethical implications are fully studied and addressed.

The vast majority of today's AI systems are data hungry and therefore need large amounts of data to train and operate them. Some ethical issues may come up during the SDLC if there is a need to address data privacy and protection issues properly. The Data must only be gathered, stored, and used in a manner that is permitted by the law. Some of the required data management practices include getting informed consent and handling of sensitive data.

As AI systems become increasingly independent, they may make judgments and perform acts with serious implications. Identifying who is responsible and accountable for system behaviors is a serious ethical issue.

To maintain ethical norms and ensure the responsible use of these technologies, effective processes for assigning responsibility, establishing accountability, and mitigating possible hazards connected with autonomy must be put in place.

If AI systems are not carefully managed, they have the potential to worsen existing inequities in society. Ethical implications should be considered to ensure fair access to and benefits from AI technologies while avoiding increasing inequities based on socioeconomic position, ethnicity, gender, and other protected traits. Efforts should be made to prevent the establishment of new digital divides while also promoting equitable access and advantages for all.

Ethical concerns occur when AI systems make judgments with ethical implications, such as in cases of safety, security, or social effects. We must ensure that ethical decision-making processes are integrated into the SDLC of AI technologies.

AI systems may have unforeseen outcomes that are difficult to anticipate. When unintended effects cause harm, unwanted uses, or unanticipated societal ramifications, ethical concerns may develop. Anticipating and addressing potential unintended outcomes of AI technology within the SDLC process is critical for reducing risks and maintaining ethical standards.

Careful assessment of these ethical implications, as well as proactive measures to resolve them, is critical for ensuring responsible and ethical development and use of AI technology in SDLC processes.

Consistency and Trust Matter

AI models are not flawless and may have limits. To minimize potential difficulties such as inaccurate findings, system breakdowns, or loss of user trust, it is critical to rigorously test the reliability and trustworthiness of AI models used in SDLC. AI models should be rigorously tested, validated, and monitored to ensure accuracy, reliability, and safety.

AI systems must be credible, performing consistently and accurately in a variety of settings and scenarios. Unreliable AI systems might produce incorrect results, misinterpretations, or unanticipated actions, which can have serious ramifications in a variety of disciplines, including safety-critical applications. To guarantee that AI systems are adequately tested, validated, and certified for reliability throughout the SDLC process, trust and responsible deployment are essential.

Trustworthiness is a very important factor that can determine the various rate of adoption. The system should be designed in a way that users can easily understand how it works, the system should be held accountable for its actions, the system should be as predictable as possible, and users should be able to influence the system and its effects.

Ensuring that AI systems are adequately evaluated and verified during the SDLC process is critical for establishing their dependability and trustworthiness. This includes reviewing the system's performance against set metrics, verifying its accuracy, detecting any biases, and assessing its robustness under various scenarios. To ensure that AI systems produce reliable and trustworthy results, a rigorous validation and verification process should be implemented.

Implementing a strong quality assurance procedure in the SDLC of AI technologies is critical to ensuring their dependability and trustworthiness. This includes adhering to software engineering best practices, following established code standards, testing thoroughly, and conducting rigorous reviews and audits. A quality assurance strategy should be in place to discover and resolve potential flaws, vulnerabilities, or biases in AI systems.

Monitoring and maintaining AI systems throughout their lifecycle is also crucial. Regular monitoring should be carried out to discover and address any performance issues, biases, or unintended consequences that may occur during system operation. Maintenance actions such as updates, patches, and improvements should be performed to guarantee that the system remains dependable, accurate, and trustworthy over time.

Establishing trust requires ensuring transparency and accountability in the development and usage of AI. This involves clearly documenting the system's design, operation, and decision-making processes, as well as establishing clear lines of accountability for the system's actions. Transparently explaining the AI's capabilities, limitations, and potential hazards to stakeholders and users is critical for fostering confidence and responsible use.

Human Oversight

While AI can produce code fast and effectively, it may require more creativity and human developers to approach challenges and answers in novel and imaginative ways. Its coding might generate code with flaws and bugs that are difficult to discover and resolve, demanding extensive human intervention and skill to correct. Many other aspects require human oversight.

There are many important ethical factors in the SDLC when adopting AI technologies. There is widespread anxiety that if AI learns to write code, software developers will become obsolete. However, software development is difficult and requires a human brain to guide it. While AI can automate many portions of the SDLC, human oversight is still required. Human interaction and decision-making are crucial for guaranteeing software development process quality, security, and appropriateness. With greater reliance on AI, proper human supervision is required to minimize unforeseen outcomes and hazards. Humans should have ultimate decision-making authority when building and operating AI systems.

While AI technologies can provide useful insights and recommendations, it is ultimately up to human stakeholders to make critical decisions. This involves judgments on system architecture, training data, performance measures, and deployment tactics.

Human oversight should be employed to ensure that ethical principles are observed during the development and deployment of AI systems. Ethical considerations such as fairness, bias, privacy, openness, and responsibility must be properly addressed at each stage of the SDLC. Human stakeholders should actively examine and guide the ethical implications of AI technology to ensure their responsible and ethical use.

We should continuously monitor and validate the performance and behavior of AI systems, watching out for biases, errors, and unexpected effects in system outputs, as well as confirming the system's accuracy, reliability, and fairness. Human stakeholders should always confirm that an AI system's performance is consistent with stated aims and ethical requirements.

Human monitoring should take into account the adaptability and flexibility of AI systems. AI technology is designed to adapt, and its behavior may change in response to new data or upgrades. Human stakeholders must be willing to adapt and alter an AI system as needed to ensure its ethical and responsible operation throughout the SDLC process.

Human stakeholders must comprehend how the AI system makes its decisions and suggestions. This will enable effective oversight, accountability, and detection of any biases or ethical issues. AI systems should be built to be interpretable and explainable, facilitating human comprehension and oversight.

User feedback and input should be encouraged when building and deploying AI systems. Feedback from users, stakeholders, and affected parties can shed light on potential biases, ethical concerns, or unforeseen repercussions. Human stakeholders should actively seek and incorporate comments to develop the AI system and guarantee it is consistent with ethical principles and stakeholder values.

Legal and Regulatory Compliance

The use of AI in SDLC may be subject to a variety of legal and regulatory restrictions, including data privacy legislation, intellectual property rights, and industry-specific rules. To avoid legal and reputational issues, ensure that AI is used in SDLC in accordance with all applicable rules and regulations.

When employing AI in SDLC, firms must consider data protection rules and regulations such as the General Data Protection Regulation (GDPR) in the European Union, the Health Insurance Portability and Accountability Act (HIPAA) in the United States, and comparable laws in other countries. These rules govern the acquisition, storage, processing, and sharing of personal information, including that used to train and deploy AI models. Organizations must ensure compliance with these regulations by acquiring necessary consent, maintaining data security, and respecting individuals' privacy rights.

Different industries may have their own norms and standards that must be followed while implementing AI in SDLC. For example, rules governing the development and use of AI systems in healthcare contexts include HIPAA and the Food and Drug Administration (FDA). Similarly, in the financial industry, rules such as the Dodd-Frank Act and Anti-Money Laundering (AML) legislation may have an impact on the usage of artificial intelligence in financial applications. Organizations must be aware of industry-specific rules, and AI can be used to assure compliance.

Organizations must address the intellectual property rights associated with the AI models, algorithms, and software that are generated or used. To minimize any legal conflicts or infringement claims, intellectual property concerns such as ownership, licensing, and usage rights must be properly reviewed and addressed.

Organizations should also consider preserving their own intellectual property rights and ensuring that any third-party AI technology used is done so in accordance with applicable IP rules and regulations.

The employment of AI in SDLC poses issues of liability and accountability. If AI systems have unexpected implications or negative outcomes, it may be difficult to pinpoint who is to blame.

Organizations must evaluate the legal consequences of incorporating AI into SDLC and ensure that adequate liability and accountability mechanisms are in place. This could include clearly defining duties and responsibilities, establishing contractual agreements, and acquiring appropriate insurance coverage to limit legal risks.

Some legal and regulatory compliance problems center on the openness and clarity of AI systems. Regulations may force enterprises to provide explanations and justifications for judgments made by AI systems, particularly in high-risk areas such as healthcare, finance, and law. Organizations must guarantee that their AI systems are visible and explainable and provide the necessary explanations to comply with applicable legislation.

Specialized Skills

Implementing AI models within the SDLC requires certain capabilities of the team. It is imperative that organizations ensure that their development teams that are responsible for implementing AI technology in the SDLC are well equipped and trained to do so appropriately and cautiously. This is because the absence of information can lead to wrong usage of AI; for instance, one can develop an inefficient AI application or even harmful one.

AI involves the creation of machine learning models and, therefore, involves data science, machine learning algorithms, and statistical knowledge. Data scientists have the responsibility of acquiring data, transforming it, creating features, and then choosing, training, evaluating as well as tuning the models in a bid to get better and reliable results.

As AI systems become more integrated into the SDLC, software development knowledge will become increasingly important. This includes skills in programming languages such as Python, R, and other applicable languages, as well as knowledge of software development methods, version control, and the continuous integration/continuous deployment (CI/CD) process. Software developers oversee the implementation and integration of AI components into the SDLC pipeline and the development of APIs, guaranteeing seamless interface with existing software systems.

It is also important to have good project management skills to implement AI into SDLC in the right manner to control the process, assign roles and responsibilities, meet the deadlines, and ensure the effective collaboration of the team that may consist of both new and old members. The project managers are the individuals who are charged with the responsibility of designing, installing, and controlling the AI system, risk control, resources, and the stakeholders. It may also require dealing with changes in the organization's structure, including those in workflow, processes, and culture. Change management professionals can help with the integration of the new AI technology ensuring that there is no resistance and also in ensuring that the right information and training is given to ensure that AI is well incorporated in the SDLC.

Once AI systems have been deployed in the SDLC, continuous monitoring and maintenance are required to assure their performance and dependability. Monitoring professionals can track the performance of AI systems, identify and resolve difficulties, and guarantee that the models continue to produce accurate and relevant results. Maintenance professionals may manage updates, upgrades, and maintenance of AI components to keep them operational and up-to-date.

Transparency

In the context of applying AI in the SDLC, transparency means the ability to understand, analyze, and discuss how AI systems work and what they do. AI systems should be in a position to provide explanations of their decisions or outputs in a way that is easy for humans to comprehend.

Explainable AI (XAI) techniques, such as rule-based systems, interpretable machine learning models, or model-agnostic methodologies like Local Interpretable Model-agnostic Explanations (LIME) or SHAPley Additive Explanations (SHAP), can help provide insights into AI systems' decision-making processes. Transparent AI systems should be well-documented, with detailed information about the design, functionality, and limitations of the AI models or algorithms employed during SDLC. This documentation should be available to key stakeholders and include information on the training data, model architecture, hyperparameters, and any other important elements that can help comprehend how the AI system works.

Transparent AI systems should keep track of where the data came from, how it was processed, and how it was transformed for training and inference. Data provenance is critical for understanding the data's quality, trustworthiness, and any biases, as well as assuring compliance

CHAPTER 3   OPPORTUNITIES AND IMPACTS OF GENAI

with data protection legislation. These systems should provide clear and understandable performance indicators that represent the correctness, dependability, and fairness of the AI models employed in SDLC.

These metrics can assist in assessing the performance of the AI systems and hence provide valuable insights into their best practices and possible improvement areas and can be employed to check and assess the performance of the AI models when integrating them into SDLC. This can include logging, monitoring, and evaluating an AI system to see how it is functioning and to identify any biases, mistakes, or issues that could occur when in use.

AI systems should have user interfaces that provide insights into the AI models' decision-making process while also allowing users to understand and analyze the AI system's outputs. User experience (UX) design should prioritize making the AI system's behavior and decision-making process transparent and accessible to users while avoiding overcomplicated or opaque interfaces.

Aside from such concerns, AI will soon be integrated into all business applications of modern software organizations. The software development process can be improved by incorporating it into as many stages as possible. AI will become necessary for developers, as it now has more importance than ever.

The SDLC landscape is rapidly evolving. To keep ahead of the competition, organizations must be aware of emerging technologies and implement them as quickly as possible.

The advantages of AI are not restricted to the development process. The most essential component of AI is its ability to minimize time spent on any process, which is critical for enterprises. AI exists to facilitate the human experience while minimizing work. It is already making a difference and generating waves in a variety of sectors, and software development is only one example of where it may eventually have a significant impact. As developers, QA specialists, and project managers improve their productivity, firms are more likely to deliver higher-quality software.

125

CHAPTER 3   OPPORTUNITIES AND IMPACTS OF GENAI

There are always some risks that cannot be avoided. There are unrealistic expectations and unanticipated consequences. While AI improves the efficiency and productivity of software development, there are hazards associated with employing new technology that we do not fully understand, and we are all learning together.

In this chapter, many aspects of the vision of generative AI in a new commerce were introduced, in addition to disruptions brought by the pandemic on a global scale, how GenAI has impacted and accelerated the software development life cycle, the rise of the new field of prompt engineering as LLMs becomes more present, and some important aspects related to limitations and legal concerns. In the next chapter, we will see how AI is being adopted through different platforms as we explore Amazon Bedrock.

CHAPTER 4

Getting Started with Amazon Bedrock

To start our journey, we need to prepare our environment and then follow the instructions in this chapter.

Log in to AWS Console

Let's start with the login for an AWS account.

Step 1: Provide the user email address and click **Next**. If you don't have an AWS account, click the **Create a new AWS account** button.

CHAPTER 4 GETTING STARTED WITH AMAZON BEDROCK

Step 2: Provide the password and click **To enter**.

CHAPTER 4   GETTING STARTED WITH AMAZON BEDROCK

Step 3: Upon successful login, you should see the AWS console.

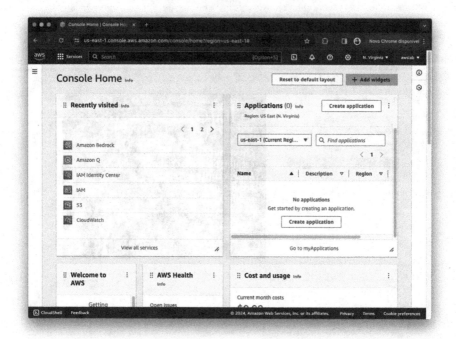

CHAPTER 4   GETTING STARTED WITH AMAZON BEDROCK

Step 4: In the search bar, type "Amazon Bedrock" and click the first result.

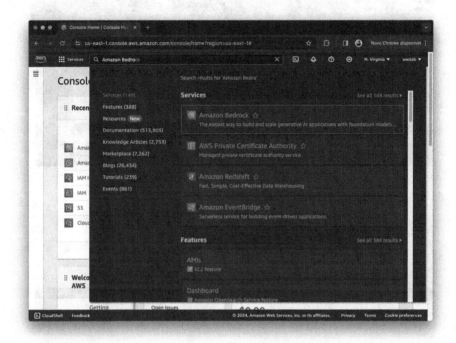

131

CHAPTER 4   GETTING STARTED WITH AMAZON BEDROCK

Step 5: When you see this window, just click the **Get started** button.

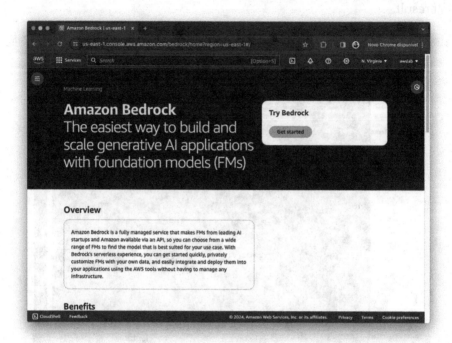

CHAPTER 4   GETTING STARTED WITH AMAZON BEDROCK

Step 6: Now, you should see the Bedrock overview page.

Working with Chat Playground

Amazon Bedrock has a playground area where you can interact with GenAI use case examples and experiment with capabilities like summarization, Q&A, and image generation supported by different foundation models (FMs).

133

CHAPTER 4   GETTING STARTED WITH AMAZON BEDROCK

Now we will explore a little bit in the playground area to get familiar with FM models and Amazon Bedrock in a practical way. Let's warm up!

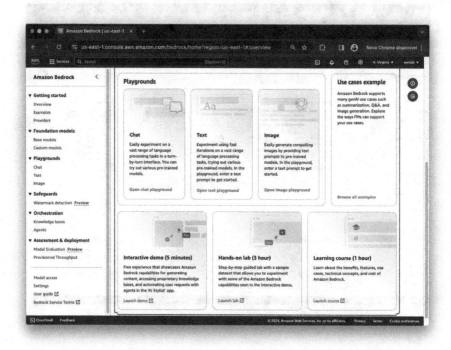

Step 1: Click **Open chat playground.**

The Chat playground will be shown with four sample use cases based on different models.

CHAPTER 4 GETTING STARTED WITH AMAZON BEDROCK

Step 2: Click the **Select Model** button.

Step 3: A pop-up window will be shown with available model providers. Click **Amazon**.

CHAPTER 4   GETTING STARTED WITH AMAZON BEDROCK

Step 4: Click **Request access link**.

We can see that Titan Text G1 – Express is not available, so we must request access.

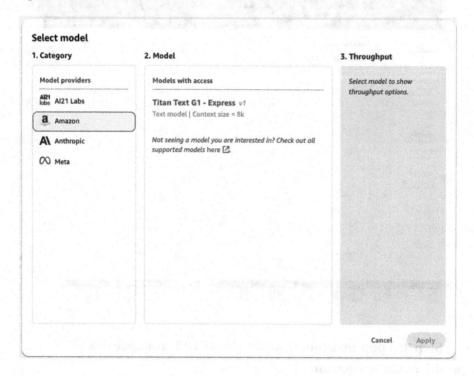

A new window will open with the Model access page where you need to request access to the related model, Amazon Titan Text G1 – Express.

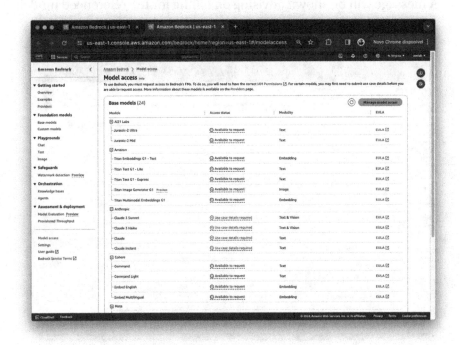

CHAPTER 4   GETTING STARTED WITH AMAZON BEDROCK

Step 5: Click **Available to request link** on the same line as the selected model.

A message will be shown advising on billing for the referenced model.

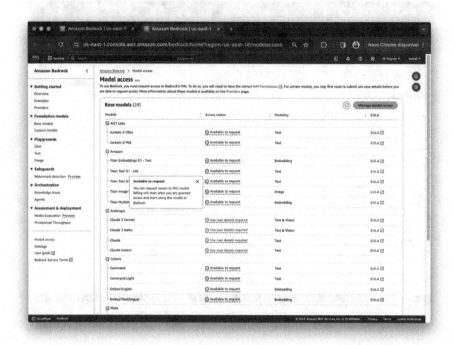

CHAPTER 4 GETTING STARTED WITH AMAZON BEDROCK

Step 6: Click the **Manage model access** button.

The model access page will make all the models available to be selected.

CHAPTER 4   GETTING STARTED WITH AMAZON BEDROCK

Step 7: Select only the Amazon Titan G1 – Express model.

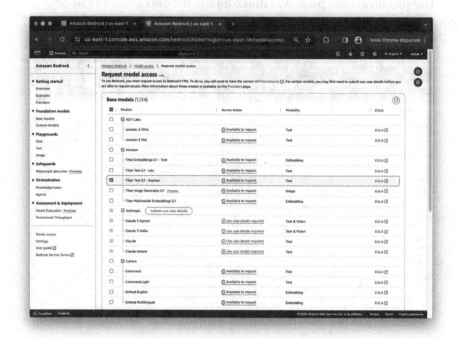

CHAPTER 4   GETTING STARTED WITH AMAZON BEDROCK

Step 8: Scroll down the model access page and click the **Request model access** button.

Now you should see that model access was granted.

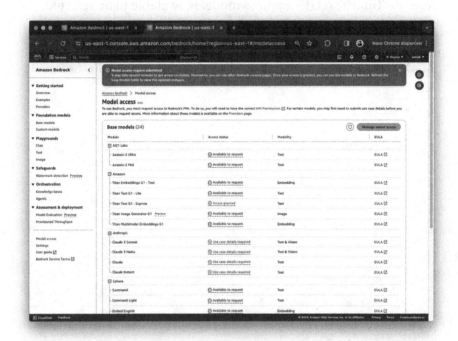

CHAPTER 4   GETTING STARTED WITH AMAZON BEDROCK

Step 9: Close this window.

You should have only the chat playground window with the select model pop-up window open.

Now the Titan Text G1 – Express model is available to be selected.

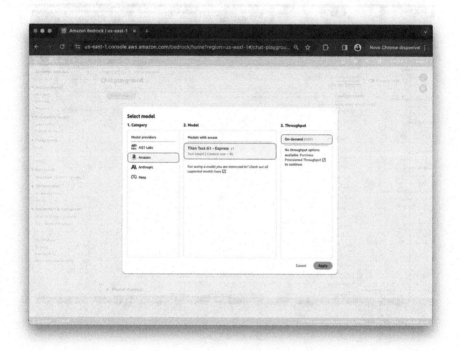

CHAPTER 4   GETTING STARTED WITH AMAZON BEDROCK

Step 10: Click the **Apply** button.

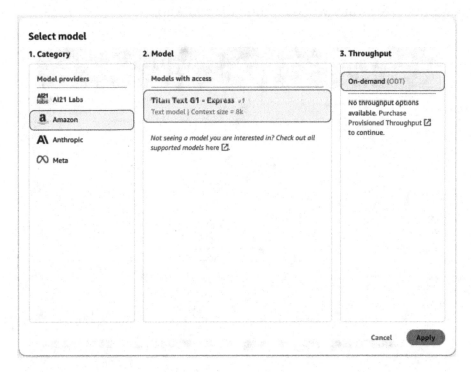

143

CHAPTER 4　GETTING STARTED WITH AMAZON BEDROCK

Once the model access is granted and you have applied it to the chat playground, the window should show the following:

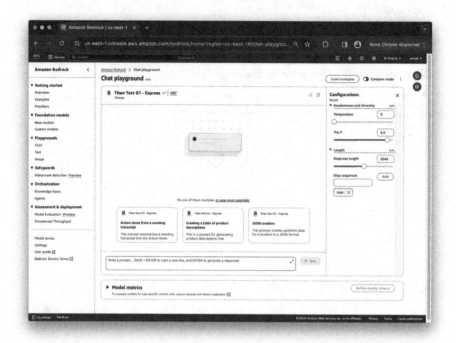

CHAPTER 4   GETTING STARTED WITH AMAZON BEDROCK

Step 11: Now, you can enter a prompt in the space below and then click the **Run** button.

Note In this example, the prompt used was "Create a action items plan to realize a wedding 25th birthday celebration event of Simpsons family." You can see that even with grammatical errors, GenAI understands the request.

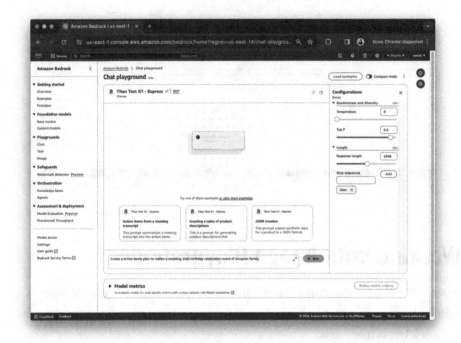

145

CHAPTER 4   GETTING STARTED WITH AMAZON BEDROCK

You can see the output generated by Amazon Bedrock using an Amazon Titan Text G1 – Express model in the following image:

Working with Image Playgrounds

Now, we will use Amazon Bedrock image playgrounds to generate an image through the Stable Diffusion FM model. Let's start from the Amazon Bedrock overview page.

CHAPTER 4 GETTING STARTED WITH AMAZON BEDROCK

Step 1: Click **Open image playground** link.

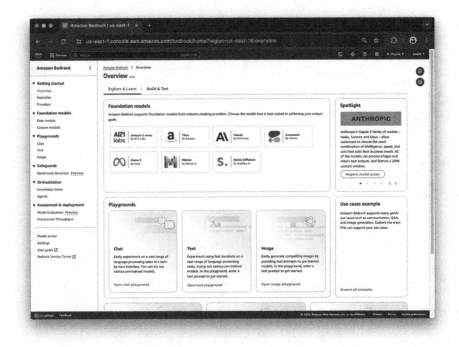

147

CHAPTER 4   GETTING STARTED WITH AMAZON BEDROCK

Step 2: Click the **Create an image** button.

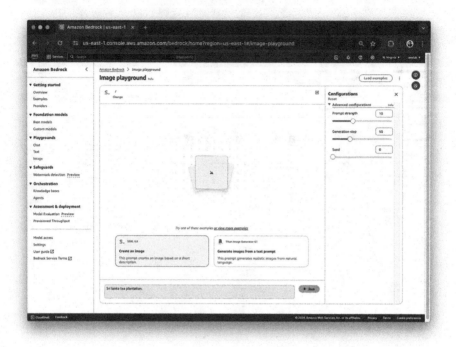

Step 3: After clicking the **Create an image** button, enter a short prompt: "Sri lanka tea plantation."

CHAPTER 4   GETTING STARTED WITH AMAZON BEDROCK

Step 4: Click the **Run** button.

You should see a red warning message indicating that you don't have access to the Stable Diffusion model.

CHAPTER 4   GETTING STARTED WITH AMAZON BEDROCK

Step 5: Click the **Change** link right below the Stable Diffusion logo.

You should see a pop-up window with the model to be selected. You haven't been granted model access yet, so let's fix that.

CHAPTER 4   GETTING STARTED WITH AMAZON BEDROCK

Step 6: Click the **Request access** link.

A new window will open with the model access management console.

CHAPTER 4   GETTING STARTED WITH AMAZON BEDROCK

Step 7: Scroll down the page, select the two AI Stable Diffusion models, and then click the **Save changes** button.

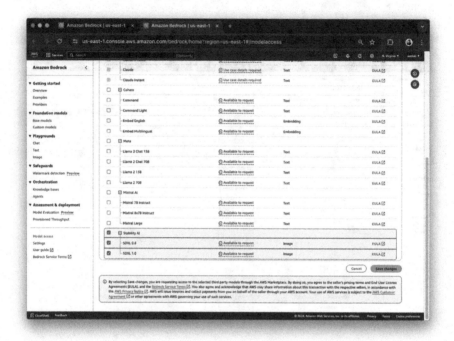

CHAPTER 4   GETTING STARTED WITH AMAZON BEDROCK

You can see that Amazon Bedrock is enabling access to the model, so don't worry if the status shows that access granting is in progress.

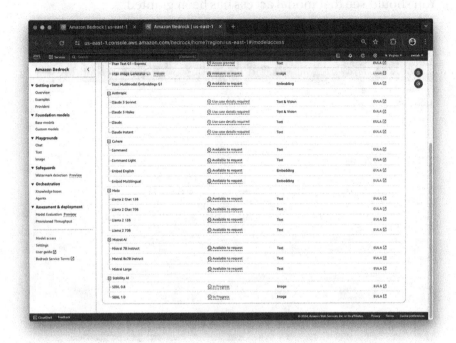

CHAPTER 4   GETTING STARTED WITH AMAZON BEDROCK

Step 8: After approximately two to three minutes, click **Reload** to reload your page.

You should see that model access has been granted.

CHAPTER 4   GETTING STARTED WITH AMAZON BEDROCK

Step 9: Close this window.

You should have only the chat playground window with the select model pop-up window open. Now the Stable Diffusion model is available to be selected.

CHAPTER 4   GETTING STARTED WITH AMAZON BEDROCK

Step 10: Click the **Apply** button.

The window has now returned to the image playground with a prompt enabled.

CHAPTER 4   GETTING STARTED WITH AMAZON BEDROCK

Step 11: Click the **Run** button.

After a couple of seconds, an AI-generated image will come up on your screen, as shown in the following example.

In this example, the prompt used was "Sri lanka tea plantation."

Step 12 (Optional): Change the prompt to "NYC times square."

157

CHAPTER 4   GETTING STARTED WITH AMAZON BEDROCK

Step 13 (Optional): Click the **Run** button.

After a couple of seconds, an AI-generated image will come up on your screen, as shown in the following example:

Step 14 (Optional): Click the generated image.

Step 15 (Optional): Click the **Download image** button.

Working with Image Playground (Advanced)

Now we will use the Amazon Bedrock image playground to change an image by prompting via the Amazon Titan Stable Diffusion FM model. Let's start from the Amazon Bedrock overview page.

CHAPTER 4   GETTING STARTED WITH AMAZON BEDROCK

Step 1: Click the **Browse all examples** link.

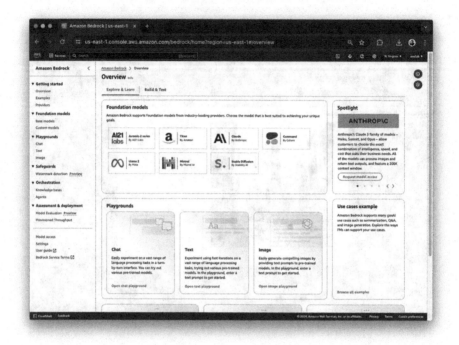

Step 2: Click the **Browse all examples** link.

You should see the Examples page listing several different use cases and respective models.

160

CHAPTER 4   GETTING STARTED WITH AMAZON BEDROCK

Step 3: Type the word "replace" in the input field, as shown in the following image:

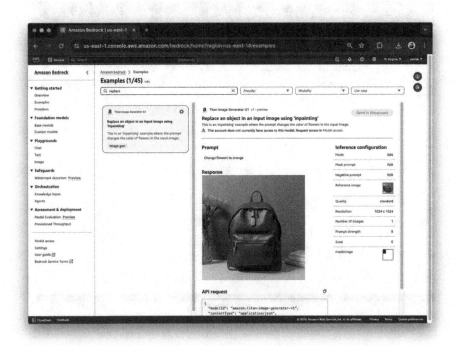

Note that you don't have access to this model.

CHAPTER 4   GETTING STARTED WITH AMAZON BEDROCK

Step 4: Click the **Model access** link.

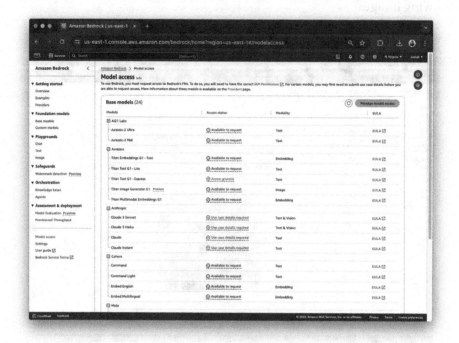

CHAPTER 4   GETTING STARTED WITH AMAZON BEDROCK

Step 5: Click the **Manage model access** button.

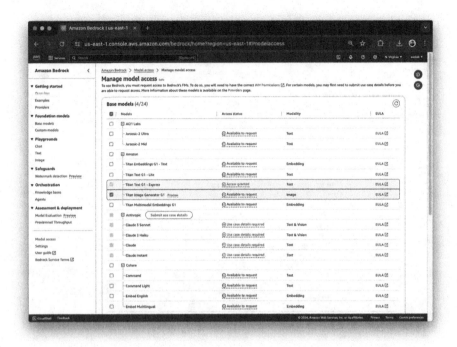

CHAPTER 4   GETTING STARTED WITH AMAZON BEDROCK

Step 6: Select the Titan Image Generator G1 model.

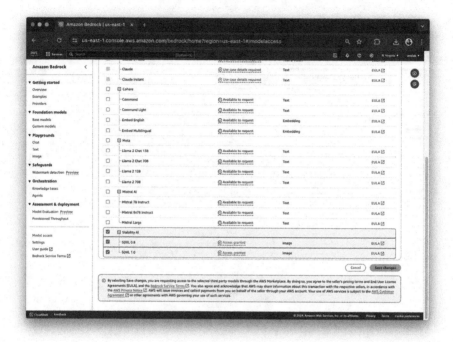

CHAPTER 4   GETTING STARTED WITH AMAZON BEDROCK

Step 7: Click the **Save change** button at the bottom of this page.

You should have access to Titan Image Generator G1 now, as shown in the following:

CHAPTER 4   GETTING STARTED WITH AMAZON BEDROCK

Step 8: Click the **Examples** option to come back to the examples page.

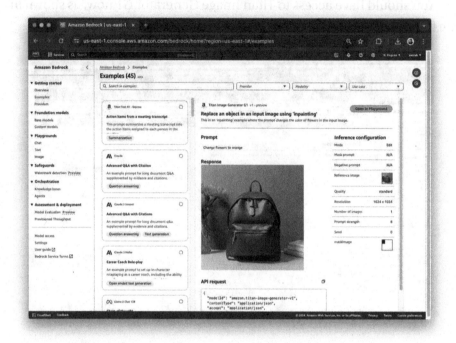

CHAPTER 4   GETTING STARTED WITH AMAZON BEDROCK

Step 9: Click the **Open in Playground** button.

You should see the image loaded in the playground area.

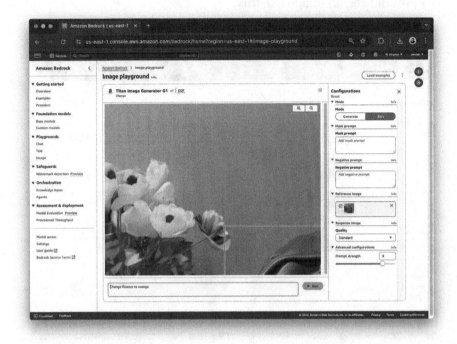

CHAPTER 4   GETTING STARTED WITH AMAZON BEDROCK

Step 10: Click the **magnifying glass** icon 13 times to see the entire image.

You should see an image like this:

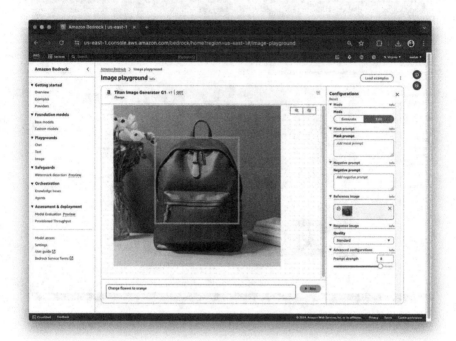

CHAPTER 4 GETTING STARTED WITH AMAZON BEDROCK

Step 11: Move the square over the flower.

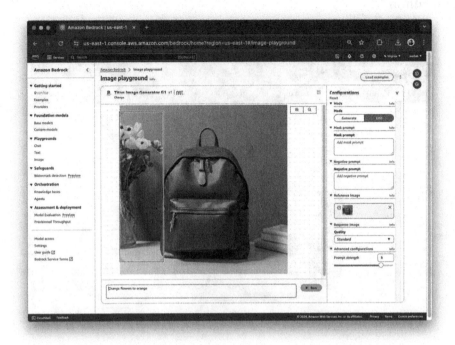

CHAPTER 4   GETTING STARTED WITH AMAZON BEDROCK

Step 12: Enter the prompt "Change flowers to orange" and click the **Run** button.

The change process will begin.

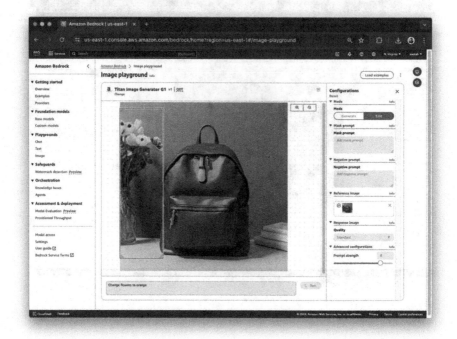

After a couple of seconds, you will see that the color of the flowers has changed to orange, following the prompt you entered:

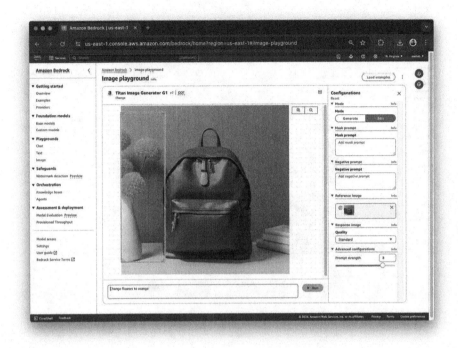

Step 13: Now change the prompt to "Change flowers to blue."

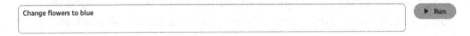

CHAPTER 4 GETTING STARTED WITH AMAZON BEDROCK

Step 14: Click the **Run** button.

You can see that the image has changed according to the prompt.

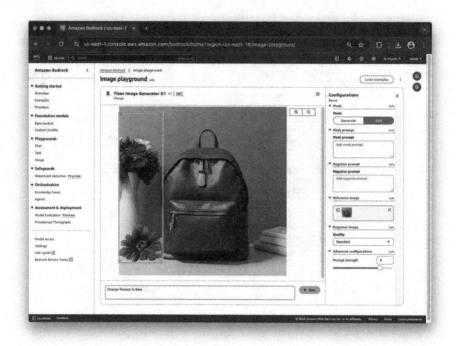

In this chapter, you learned how to take advantage of Amazon Bedrock to generate text and images supported by the Titan Text G1 – Express, Stable Diffusion, and Titan Image Generator G1 foundation models. You also briefly explored the playground area. In the next chapter, we will keep practicing using Amazon Bedrock with more diverse use cases.

CHAPTER 5

Getting Started with GenAI Using SAP BTP and Amazon Bedrock

In this chapter, we will explore integrate a GenAI large language model (LLM) into Amazon SageMaker Notebook to improve the user's experience using business data stored in SAP HANA Cloud. We will use Amazon Bedrock, the Python language, and Langchain.

CHAPTER 5   GETTING STARTED WITH GENAI USING SAP BTP AND AMAZON BEDROCK

To start our journey, we need to prepare our environment. Follow these instructions:

Step 1: Create your SAP BTP Trial account.

URL: `https://account.hanatrial.ondemand.com/`

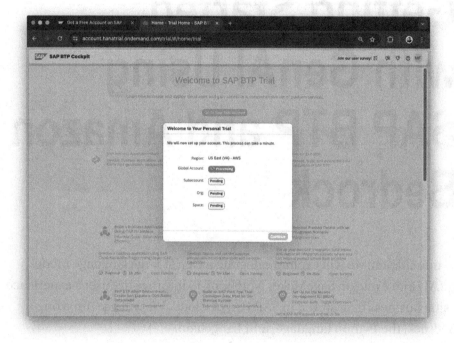

CHAPTER 5 GETTING STARTED WITH GENAI USING SAP BTP AND AMAZON BEDROCK

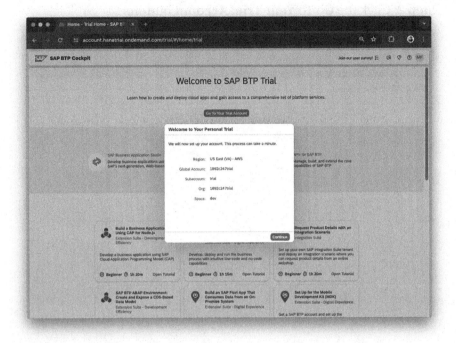

CHAPTER 5   GETTING STARTED WITH GENAI USING SAP BTP AND AMAZON BEDROCK

Step 2: Click the **Go To Your Trial Account** button.

CHAPTER 5 GETTING STARTED WITH GENAI USING SAP BTP AND AMAZON BEDROCK

Now, you can see the SAP BTP Cockpit for your trial account.

CHAPTER 5   GETTING STARTED WITH GENAI USING SAP BTP AND AMAZON BEDROCK

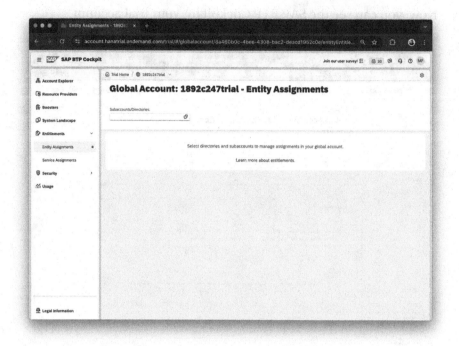

Step 3: Click **Entity Assignments** on the left side of the SAP BTP Cockpit.

CHAPTER 5 GETTING STARTED WITH GENAI USING SAP BTP AND AMAZON BEDROCK

Step 4: Select your trial account in the Subaccounts/Directories list.

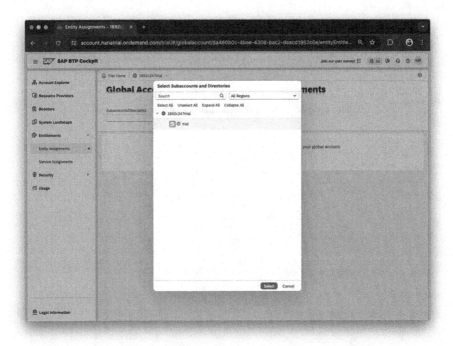

CHAPTER 5   GETTING STARTED WITH GENAI USING SAP BTP AND AMAZON BEDROCK

Step 5: Type "HANA" in the search field to see if that service is available.

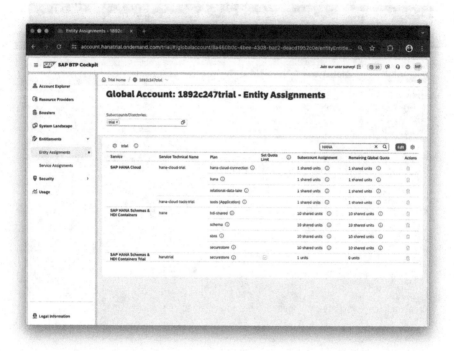

CHAPTER 5   GETTING STARTED WITH GENAI USING SAP BTP AND AMAZON BEDROCK

Step 6: Click the trial account name to see the Overview page.

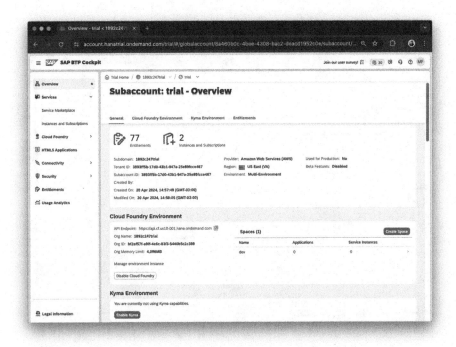

CHAPTER 5   GETTING STARTED WITH GENAI USING SAP BTP AND AMAZON BEDROCK

Step 7: Click the ellipsis to create a SAP HANA Cloud instance.

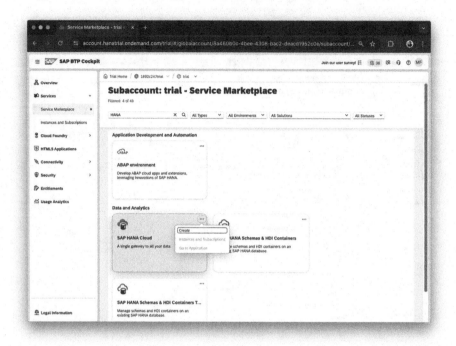

CHAPTER 5   GETTING STARTED WITH GENAI USING SAP BTP AND AMAZON BEDROCK

Step 8: Select the **tools** plan.

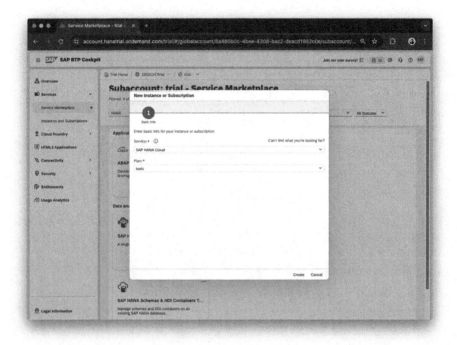

CHAPTER 5   GETTING STARTED WITH GENAI USING SAP BTP AND AMAZON BEDROCK

Step 9: Click **Security** and then **Users** to check the user rights.

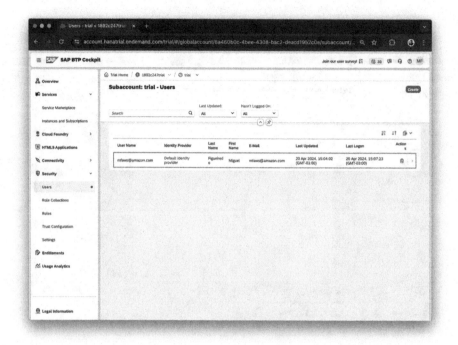

CHAPTER 5 GETTING STARTED WITH GENAI USING SAP BTP AND AMAZON BEDROCK

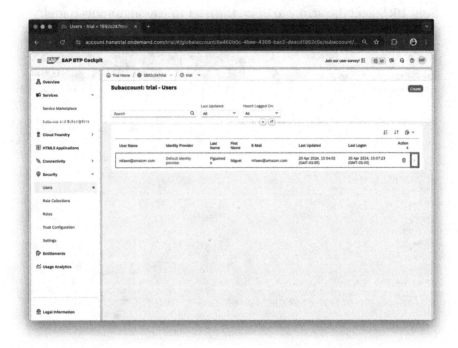

CHAPTER 5   GETTING STARTED WITH GENAI USING SAP BTP AND AMAZON BEDROCK

Step 10: Select the arrow to open the user role assignments panel.

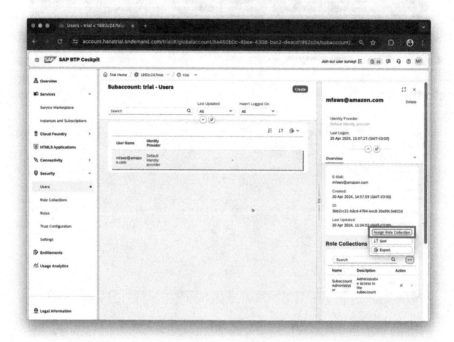

CHAPTER 5   GETTING STARTED WITH GENAI USING SAP BTP AND AMAZON BEDROCK

Step 11: Click **Assign Role Collection.**

Step 12: Type "HANA" in the search field.

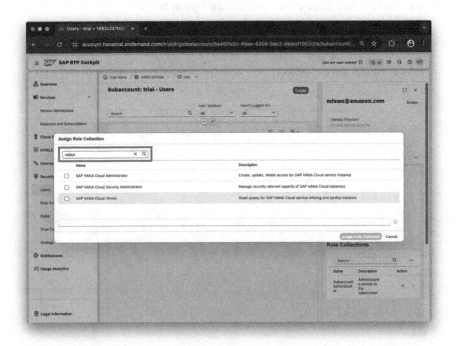

CHAPTER 5   GETTING STARTED WITH GENAI USING SAP BTP AND AMAZON BEDROCK

Step 13: Select **SAP HANA Administrator** and click the **Assign Role Collection** button.

Now you can see that you have the assigned role.

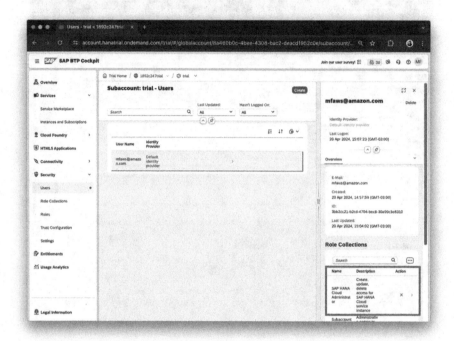

CHAPTER 5   GETTING STARTED WITH GENAI USING SAP BTP AND AMAZON BEDROCK

Step 14: Navigate to **Instances and Subscriptions** using the left menu and then click **SAP HANA Cloud**.

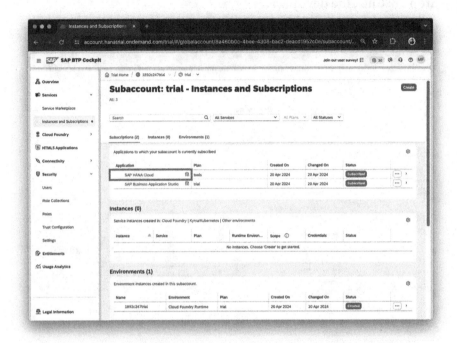

CHAPTER 5   GETTING STARTED WITH GENAI USING SAP BTP AND AMAZON BEDROCK

Step 15: Click the trial account name on the top header to see the Overview page.

Step 16: Click the dev space.

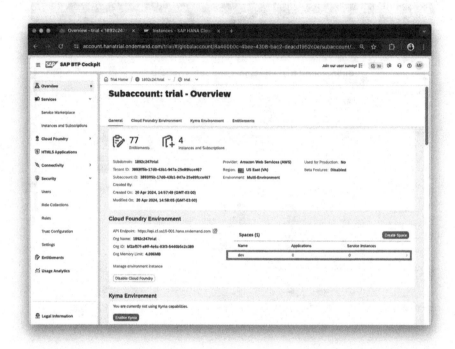

CHAPTER 5   GETTING STARTED WITH GENAI USING SAP BTP AND AMAZON BEDROCK

Step 17: Click **SAP HANA Cloud**.

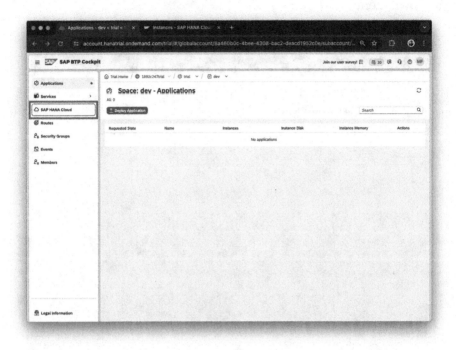

CHAPTER 5   GETTING STARTED WITH GENAI USING SAP BTP AND AMAZON BEDROCK

Step 18: Click the **Create** button and then **SAP HANA Database** in the drop-down menu.

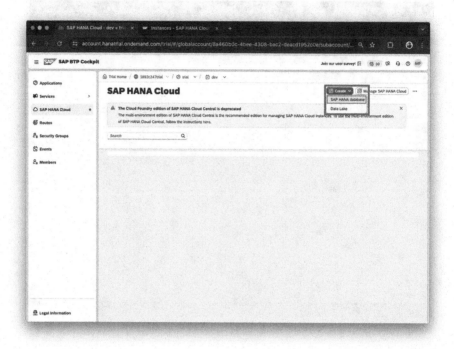

CHAPTER 5   GETTING STARTED WITH GENAI USING SAP BTP AND AMAZON BEDROCK

Step 19: Click the **Sign in with default identity provider** button.

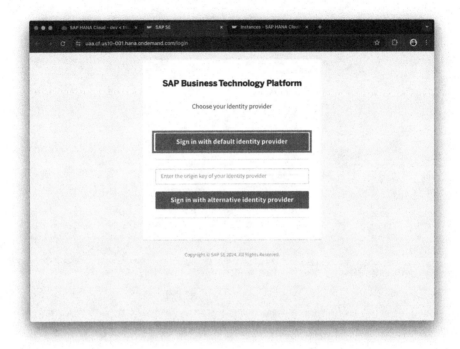

CHAPTER 5 GETTING STARTED WITH GENAI USING SAP BTP AND AMAZON BEDROCK

Step 20: Provide the user email and password.

194

CHAPTER 5   GETTING STARTED WITH GENAI USING SAP BTP AND AMAZON BEDROCK

You will see an application authorization page.

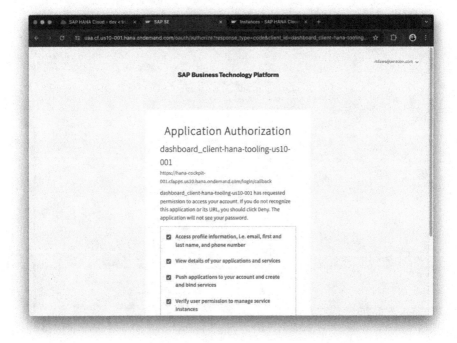

CHAPTER 5   GETTING STARTED WITH GENAI USING SAP BTP AND AMAZON BEDROCK

Step 21: Scroll down the window and click the **Authorize** button.

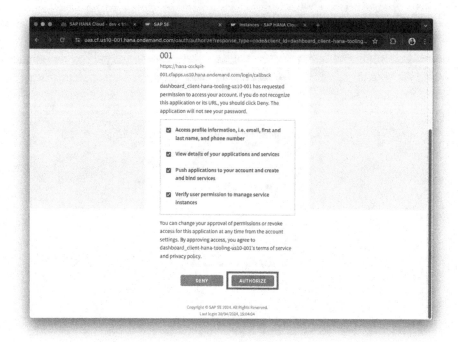

CHAPTER 5   GETTING STARTED WITH GENAI USING SAP BTP AND AMAZON BEDROCK

The SAP HANA Cloud instance manager should open.

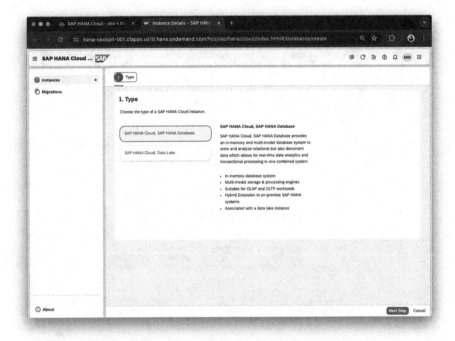

CHAPTER 5   GETTING STARTED WITH GENAI USING SAP BTP AND AMAZON BEDROCK

Step 22: Start creating your instance by providing the instance name and password.

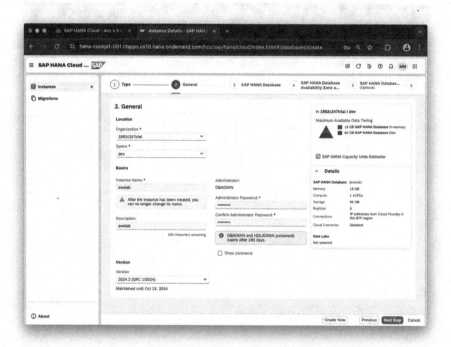

CHAPTER 5   GETTING STARTED WITH GENAI USING SAP BTP AND AMAZON BEDROCK

Step 23: Keep all the other default values and click the **Next Step** button.

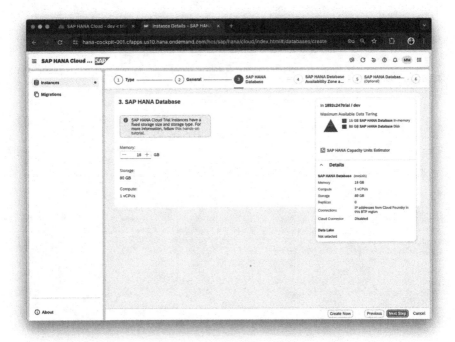

CHAPTER 5   GETTING STARTED WITH GENAI USING SAP BTP AND AMAZON BEDROCK

Step 24: Keep all the default values on this page and click the **Next Step** button.

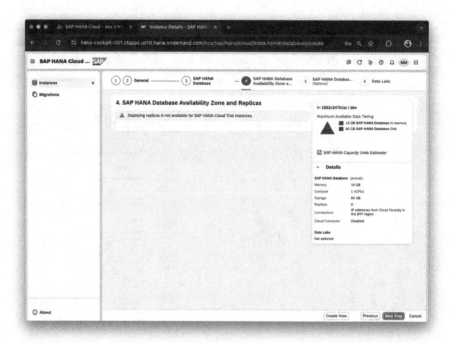

CHAPTER 5   GETTING STARTED WITH GENAI USING SAP BTP AND AMAZON BEDROCK

Step 25: Select **Allow all IP addresses** and click the **Next Step** button.

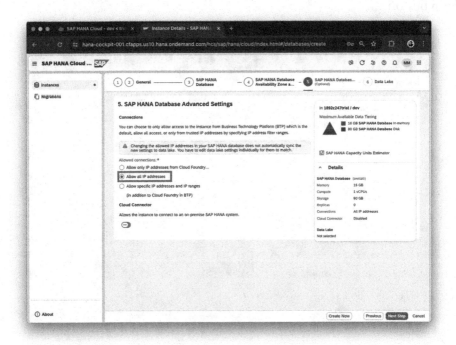

201

CHAPTER 5   GETTING STARTED WITH GENAI USING SAP BTP AND AMAZON BEDROCK

Step 26: Keep all the default values on this page and click the **Next Step** button.

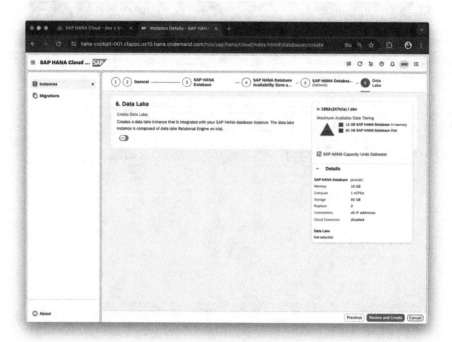

CHAPTER 5   GETTING STARTED WITH GENAI USING SAP BTP AND AMAZON BEDROCK

Step 27: Click the **Review and Create** button.

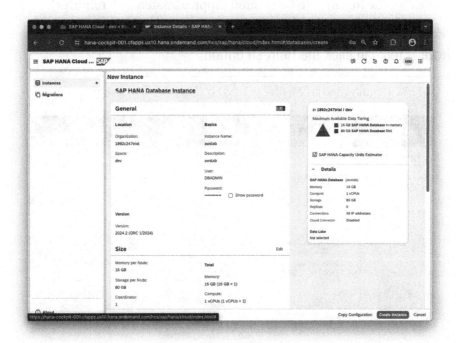

Step 28: Click the **Create Instance** button.

Wait for the instance to be created (approximately two minutes).

Note   You can click the Refresh button.

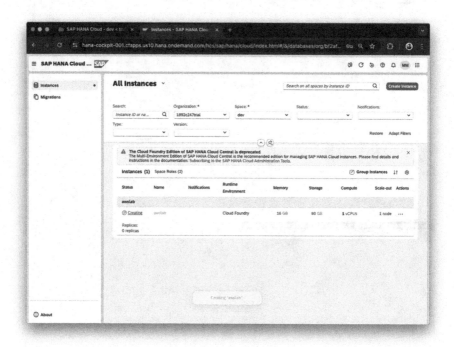

CHAPTER 5　GETTING STARTED WITH GENAI USING SAP BTP AND AMAZON BEDROCK

Now the instance is available.

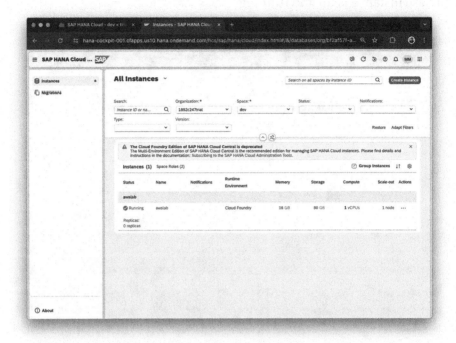

CHAPTER 5   GETTING STARTED WITH GENAI USING SAP BTP AND AMAZON BEDROCK

Step 29: Go to the SAP BTP Cockpit to see that your SAP HANA database instance is up and running.

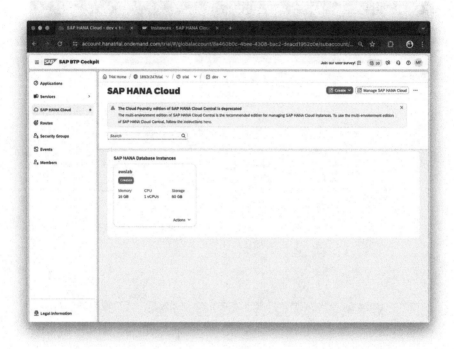

CHAPTER 5   GETTING STARTED WITH GENAI USING SAP BTP AND AMAZON BEDROCK

Step 30: Click the **Actions** link and select **Copy SQL Endpoint**.

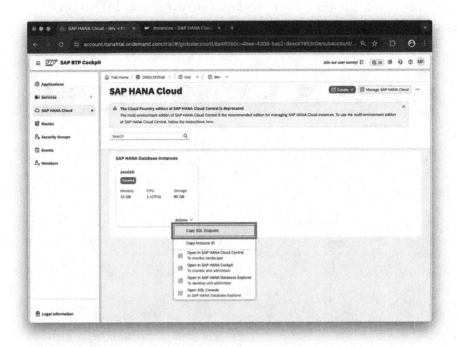

CHAPTER 5   GETTING STARTED WITH GENAI USING SAP BTP AND AMAZON BEDROCK

Step 31: Take note of the SQL endpoint. It should look like this: 73098405-c4b1-437c-89cf-3248286c20fd.hana.trial-us10.hanacloud.ondemand.com:443

Step 32: Click the **Actions** link and select **Open SQL Console**.

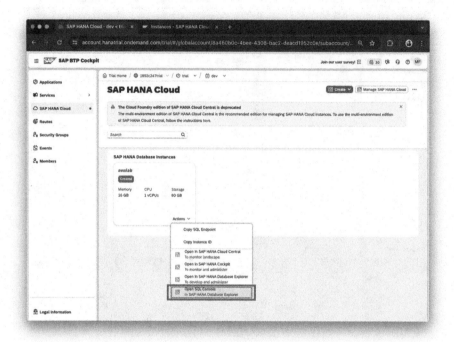

CHAPTER 5   GETTING STARTED WITH GENAI USING SAP BTP AND AMAZON BEDROCK

Now you should see a SQL console. Provide the user and password.

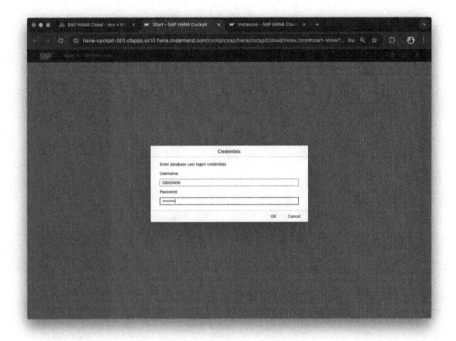

CHAPTER 5 GETTING STARTED WITH GENAI USING SAP BTP AND AMAZON BEDROCK

Now you should see a SQL console session.

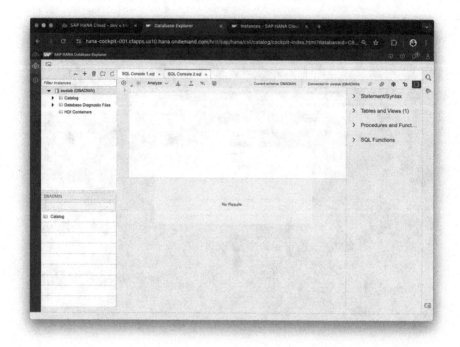

CHAPTER 5 GETTING STARTED WITH GENAI USING SAP BTP AND AMAZON BEDROCK

Step 33: Copy the SQL code block from the following file and paste it into the SQL console area:

http://link-git-hub-miguelfigueiredo/<hdb-script.sql>

Your window should look like this:

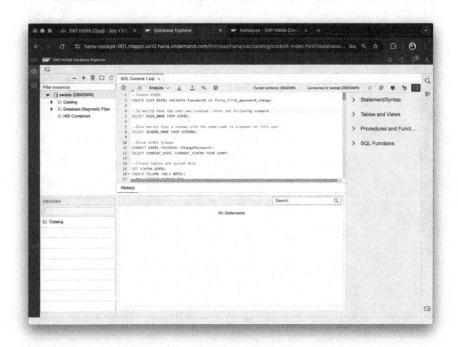

CHAPTER 5　GETTING STARTED WITH GENAI USING SAP BTP AND AMAZON BEDROCK

Step 34: Click the **Run** button.

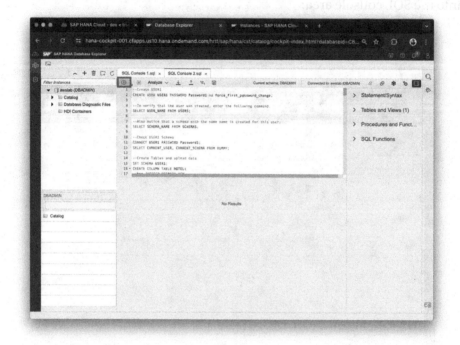

CHAPTER 5   GETTING STARTED WITH GENAI USING SAP BTP AND AMAZON BEDROCK

You should see no errors in the result.

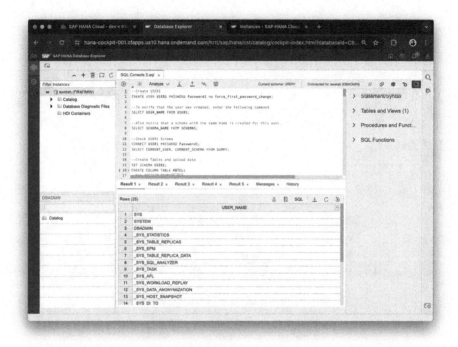

The table named "HOTEL" was created, as you can see here:

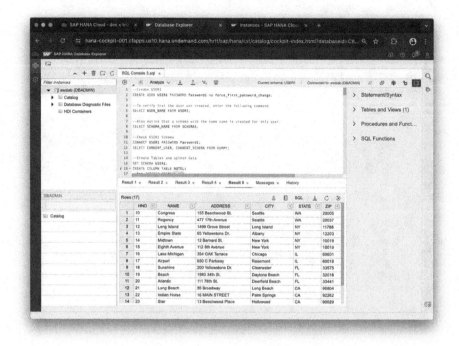

Log in to the AWS Console

Let's start by logging in to an AWS account.

CHAPTER 5   GETTING STARTED WITH GENAI USING SAP BTP AND AMAZON BEDROCK

Step 1: Provide the user email address and click the **Next** button. If you don't have an AWS account, click the **Create a new AWS account** button.

CHAPTER 5 GETTING STARTED WITH GENAI USING SAP BTP AND AMAZON BEDROCK

Step 2: Provide the password and click the **To enter** button.

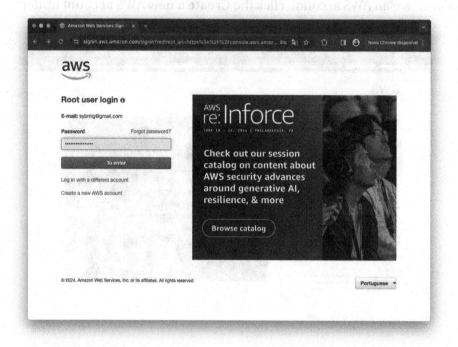

CHAPTER 5   GETTING STARTED WITH GENAI USING SAP BTP AND AMAZON BEDROCK

Step 3: Upon successfully logging in, you should see the AWS console as follows:

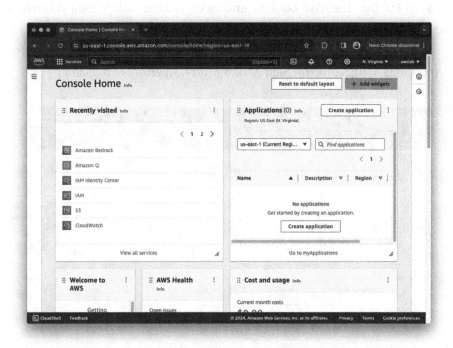

Launch Amazon SageMaker Studio

Amazon SageMaker Studio is an integrated development environment (IDE) for machine learning that enables users to create, train, debug, tune, and monitor their machine learning models. It provides all the tools needed to take models from experimentation to production while boosting productivity.

217

CHAPTER 5   GETTING STARTED WITH GENAI USING SAP BTP AND AMAZON BEDROCK

Here are the steps for onboarding to Amazon SageMaker Studio using Quick Setup:

Step 1: Open the AWS console and switch to the AWS region you would like to use.

N. Virginia is depicted in the following image. Please change the region if your lab is using Amazon SageMaker Studio in another region.

CHAPTER 5   GETTING STARTED WITH GENAI USING SAP BTP AND AMAZON BEDROCK

Step 2: If you don't see Amazon SageMaker, type "SageMaker" in the search bar and click **Amazon SageMaker.**

CHAPTER 5   GETTING STARTED WITH GENAI USING SAP BTP AND AMAZON BEDROCK

Step 3: Click **Setup for Single User.**

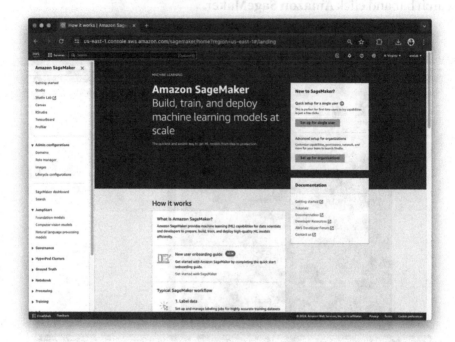

CHAPTER 5   GETTING STARTED WITH GENAI USING SAP BTP AND AMAZON BEDROCK

Amazon SageMaker will start setting up the domain. This is a one-time configuration that takes a few minutes.

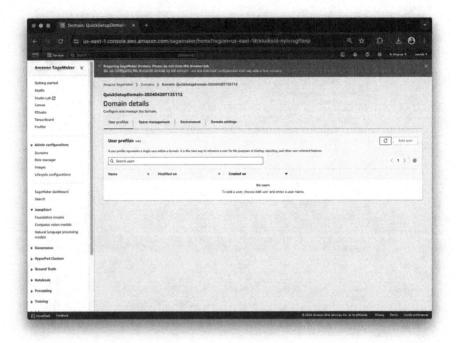

CHAPTER 5   GETTING STARTED WITH GENAI USING SAP BTP AND AMAZON BEDROCK

Once the process finishes, you should see a green message at the top of the page.

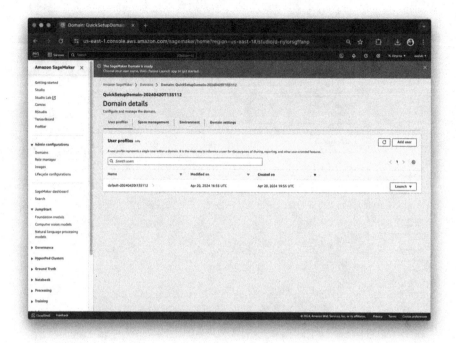

CHAPTER 5   GETTING STARTED WITH GENAI USING SAP BTP AND AMAZON BEDROCK

Step 4: Click **Domain Settings.**

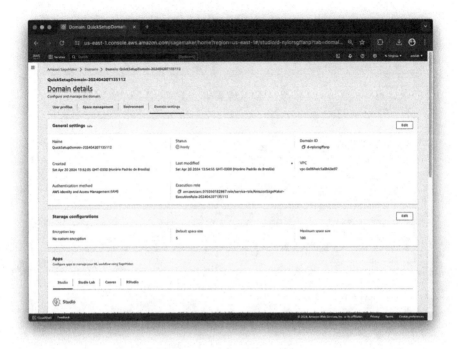

Step 5: Take note of the **Execution role** ARN.

Execution role
   arn:aws:iam::975050182987:role/service-role/AmazonSageMaker-ExecutionRole-20240420T135113

CHAPTER 5   GETTING STARTED WITH GENAI USING SAP BTP AND AMAZON BEDROCK

Step 6: Take note of the **SageMaker Execution Role** value.

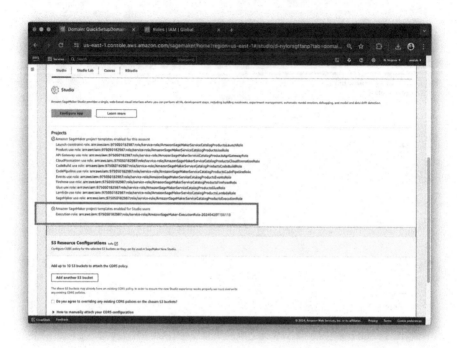

Step 7: Navigate to the Amazon IAM console available here:
`https://us-east-1.console.aws.amazon.com/iamv2/home?region=us-east-1#/roles`

CHAPTER 5   GETTING STARTED WITH GENAI USING SAP BTP AND AMAZON BEDROCK

Step 8: Search for "SageMaker-Execution Role."

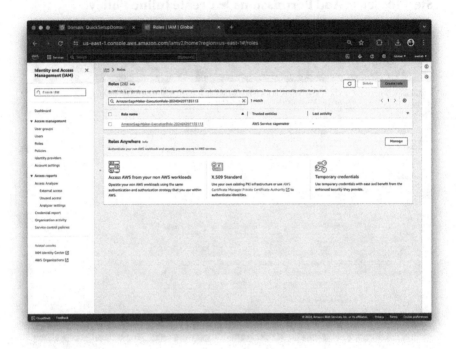

CHAPTER 5   GETTING STARTED WITH GENAI USING SAP BTP AND AMAZON BEDROCK

Step 9: Click the role name.

Step 10: Select **Add Permissions ➤ Create Inline Policy**.

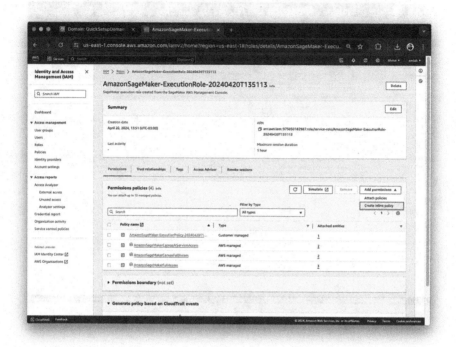

CHAPTER 5   GETTING STARTED WITH GENAI USING SAP BTP AND AMAZON BEDROCK

You should see this page:

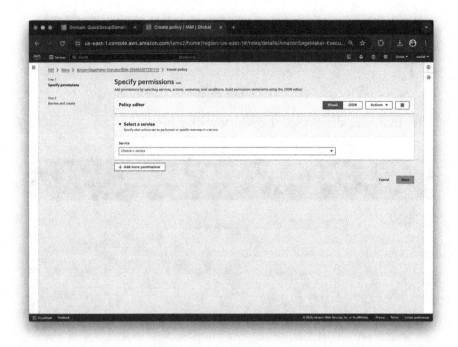

Step 11: Click the **JSON Policy Editor** button.

Step 12: Copy and paste the following code:

```
{
    "Version": "2012-10-17",
    "Statement": [
        {
            "Sid": "Statement2",
            "Effect": "Allow",
            "Action": [
                "bedrock:*"
            ],
```

CHAPTER 5 GETTING STARTED WITH GENAI USING SAP BTP AND AMAZON BEDROCK

```
            "Resource": [
                "*"
            ]
        }
    ]
}
```

After that, you should see the window like this:

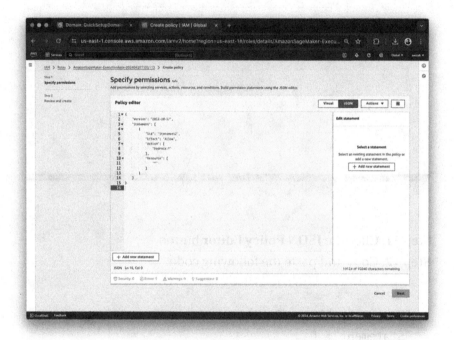

CHAPTER 5   GETTING STARTED WITH GENAI USING SAP BTP AND AMAZON BEDROCK

Step 13: Click **Next**.
Step 14: Type "**MyBedrockPolicy**" in the **Policy name** field.
Step 15: Click the **Create Policy button.**

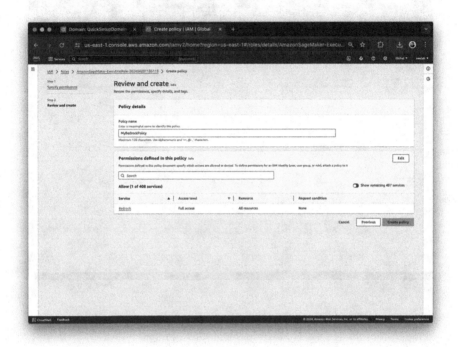

229

CHAPTER 5   GETTING STARTED WITH GENAI USING SAP BTP AND AMAZON BEDROCK

After that, the window should appear as follows:

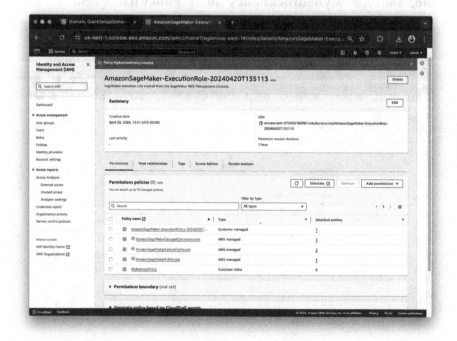

CHAPTER 5   GETTING STARTED WITH GENAI USING SAP BTP AND AMAZON BEDROCK

Step 16: Open the Amazon SageMaker console; choose the user profile, *userXX* or *defaultXX*; and then click **Open Studio**.

https://us-east-1.console.aws.amazon.com/sagemaker/home#/studio-landing

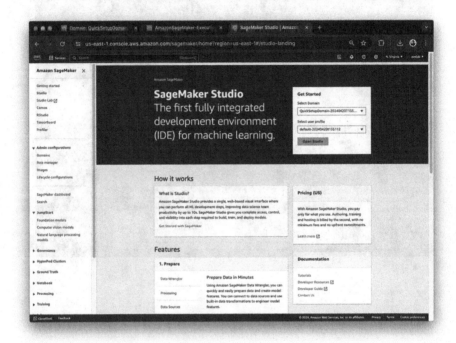

CHAPTER 5   GETTING STARTED WITH GENAI USING SAP BTP AND AMAZON BEDROCK

Step 17: Click **Open Studio**.

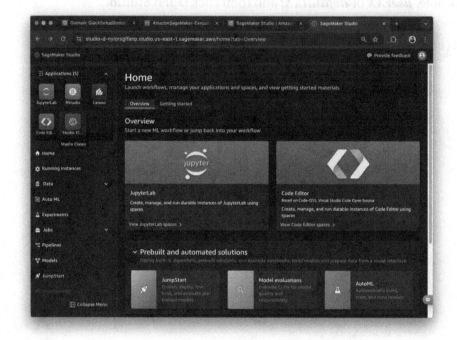

CHAPTER 5   GETTING STARTED WITH GENAI USING SAP BTP AND AMAZON BEDROCK

Step 18: Click the **Studio Classic** icon.

CHAPTER 5   GETTING STARTED WITH GENAI USING SAP BTP AND AMAZON BEDROCK

Step 19: Click the **Run** button in the Studio Classic application.

The process to start the application will begin. Wait until it gets started. You can click the **Refresh** button if necessary.

CHAPTER 5   GETTING STARTED WITH GENAI USING SAP BTP AND AMAZON BEDROCK

Once the process is finished, you'll see the action status change to Running.

CHAPTER 5   GETTING STARTED WITH GENAI USING SAP BTP AND AMAZON BEDROCK

Step 20: Click the **Open** button.

After that, a new window will open in Amazon SageMaker Studio Notebook.

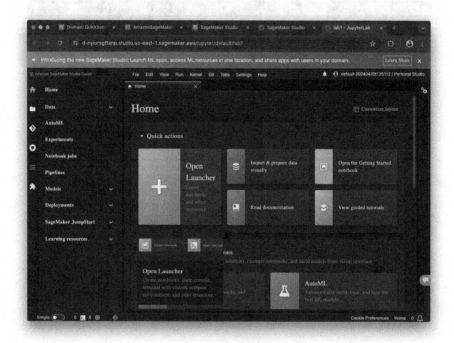

Congratulations! You have successfully launched an Amazon SageMaker Studio Notebook.

Query SAP HANA Cloud

In this section, you will learn how to perform the following tasks:

- Identify the necessary components to connect Amazon SageMaker Studio Notebook to SAP HANA Cloud, including the SAP HANA driver, credentials, and connection string.

236

CHAPTER 5   GETTING STARTED WITH GENAI USING SAP BTP AND AMAZON BEDROCK

- Use the SAP HANA Python library to establish a connection from an Amazon SageMaker Studio Notebook to SAP HANA Cloud.

- Troubleshoot common errors when connecting Amazon SageMaker Studio Notebook to SAP HANA Cloud.

Step 1: Choose **File ➤ New ➤ Notebook.**

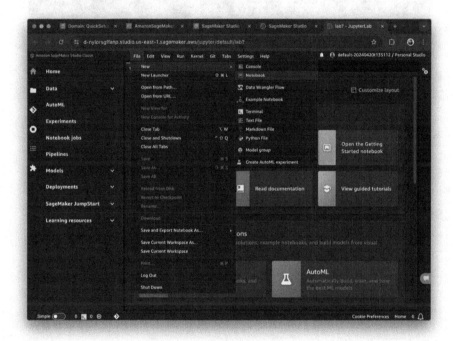

CHAPTER 5   GETTING STARTED WITH GENAI USING SAP BTP AND AMAZON BEDROCK

A new window will pop up:

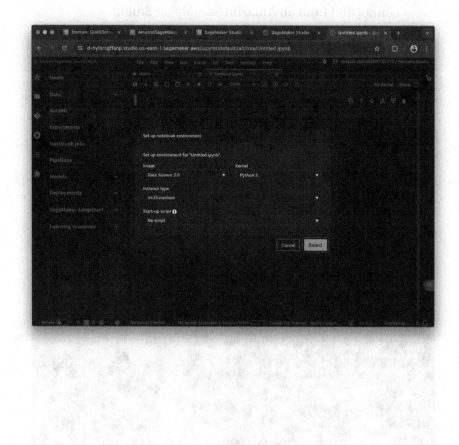

238

CHAPTER 5   GETTING STARTED WITH GENAI USING SAP BTP AND AMAZON BEDROCK

Step 2: Click the **Select** button.

Amazon SageMaker will initiate your notebook kernel.

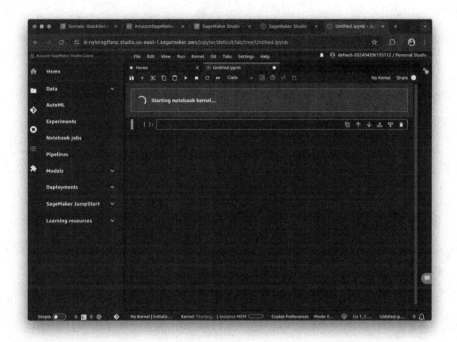

CHAPTER 5   GETTING STARTED WITH GENAI USING SAP BTP AND AMAZON BEDROCK

Step 3: Start the terminal by clicking the **$ (terminal) button** on the Amazon SageMaker Studio Notebook.

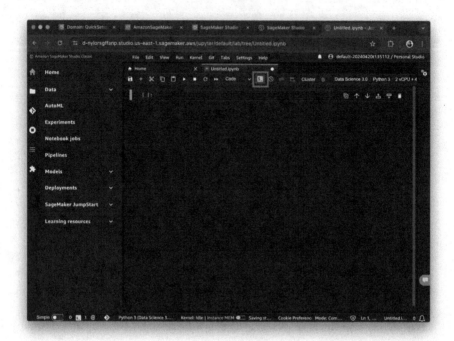

Step 4: Copy and paste the following text into the terminal session and press the Enter key:

cd /opt/conda/bin/
./pip install boto3 hdbcli

You may encounter the following error, but you can disregard this for now:

240

CHAPTER 5   GETTING STARTED WITH GENAI USING SAP BTP AND AMAZON BEDROCK

After that, you should see this result:

Step 5: Install the trusted SSL certificate for SAP HANA Cloud:

mkdir ~/.ssl
wget --no-check-certificate https://cacerts.
digicert.com/DigiCertGlobalRootCA.crt.pem -O ~/.ssl/
DigiCertGlobalRootCA.crt.pem
wget --no-check-certificate https://cacerts.digicert.
com/DigiCertSHA2SecureServerCA-2.crt.pem -O ~/.ssl/
DigiCertSHA2SecureServerCA-2.crt.pem

CHAPTER 5 GETTING STARTED WITH GENAI USING SAP BTP AND AMAZON BEDROCK

After that, you should see this result:

Go back to your Amazon SageMaker Studio Notebook, click **the plus (+) button to add cells, type the code into the cells section by section, and then** click **Play** to see the outcome for each.

For the kernel, you can choose **Data Science 3.0** with **Python 3** and **ml.t3.medium**.

CHAPTER 5 GETTING STARTED WITH GENAI USING SAP BTP AND AMAZON BEDROCK

Step 6: Change the user, password, and SQL endpoint in the following code:

```
#Import your dependencies
from hdbcli import dbapi

#Initialize your connection
conn = dbapi.connect(
    address='<changesqlendpoint.hana.trial-us10.hanacloud.ondemand.com>',
    port='443',
    user='<ChangeUser>',
    password='<ChangePassword>',
    encrypt=True,
    sslValidateCertificate=True
)
#If no errors, print connected
print('connected\n')
```

Remember the SQL endpoint you took note of in Step 31.
73098405-c4b1-437c-89cf-3248286c20fd.hana.trial-us10.hanacloud.ondemand.com:443

Tip   Remove the 443 port number from the SQL endpoint.

243

CHAPTER 5   GETTING STARTED WITH GENAI USING SAP BTP AND AMAZON BEDROCK

Click the **Play** button, and you will see the resulting message, "connected," after the code.

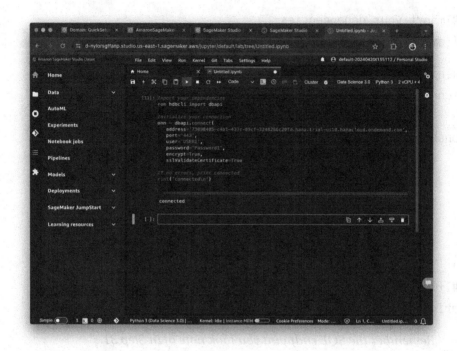

Step 7: Copy and paste the following code into a new cell:

```
cursor = conn.cursor()
sql_command = "select * from USER1.HOTEL;"
cursor.execute(sql_command)
rows = cursor.fetchall()
for row in rows:
    for col in row:
```

CHAPTER 5   GETTING STARTED WITH GENAI USING SAP BTP AND AMAZON BEDROCK

```
        print ("%s" % col, end =" ")
    print ("")
cursor.close()
conn.close()
```

Step 8: Click the **Play** button.

CHAPTER 5   GETTING STARTED WITH GENAI USING SAP BTP AND AMAZON BEDROCK

You should see the results right after the code.

CHAPTER 5   GETTING STARTED WITH GENAI USING SAP BTP AND AMAZON BEDROCK

Step 9: Now rename your Notebook in the menu, as shown in the following images:

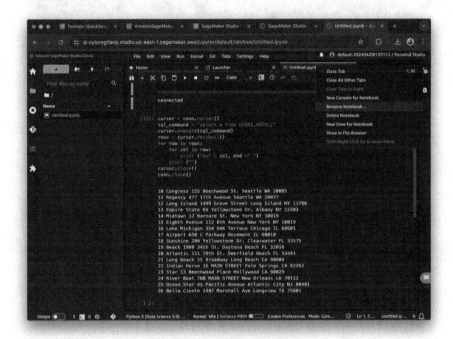

CHAPTER 5   GETTING STARTED WITH GENAI USING SAP BTP AND AMAZON BEDROCK

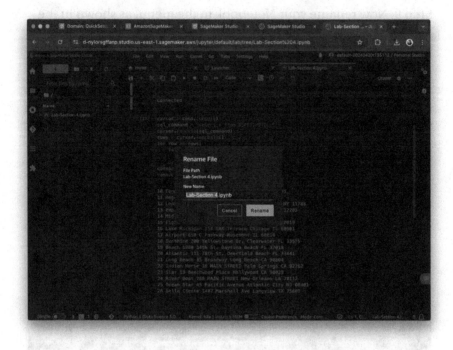

Congratulations! You have successfully concluded this section's exercise.

Integrate GenAI with SAP HANA Cloud

By the end of this section, you will be able to do the following:

- Explain the benefits of connecting Amazon SageMaker Studio Notebook to SAP HANA Cloud for accessing data with machine learning models.

- Explain Claude, an LLM capable of generating SQL queries from natural language prompts with Langchain.

248

CHAPTER 5   GETTING STARTED WITH GENAI USING SAP BTP AND AMAZON BEDROCK

- Get access to the Claude LLM into Amazon SageMaker Studio Notebook using Amazon Bedrock.

- Pass text descriptions and requests to Claude to generate equivalent SQL queries with prompt engineering.

- Connect the generated SQL queries to SAP HANA Cloud to retrieve requested data through SQLAlchemy.

- Evaluate the performance of Claude for text-to-SQL with SAP HANA Cloud data.

- Troubleshoot errors when connecting Amazon SageMaker Studio Notebook to SAP HANA Cloud and running Claude via Bedrock.

Step 1: Choose **File ➤ New Notebook.**

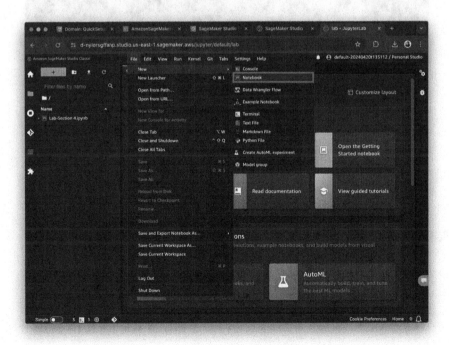

CHAPTER 5 GETTING STARTED WITH GENAI USING SAP BTP AND AMAZON BEDROCK

Step 2: Click the **Select** button.

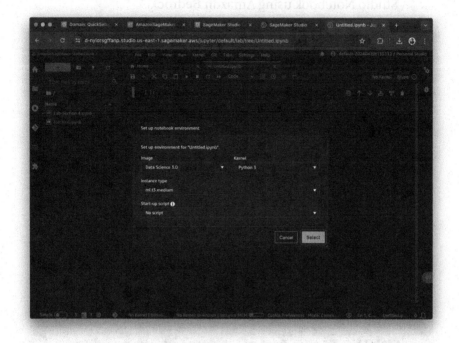

CHAPTER 5   GETTING STARTED WITH GENAI USING SAP BTP AND AMAZON BEDROCK

Step 3: Rename the notebook.

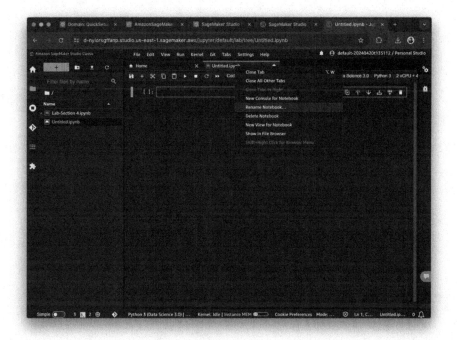

Step 4: Click the **Rename** button.

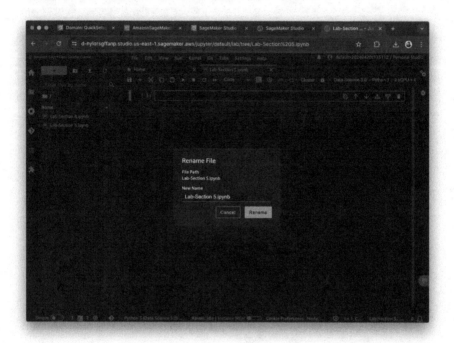

Step 5: Download the file `myutils.zip` to your computer.
URL:
https://ws-assets-prod-iad-r-iad-ed304a55c2ca1aee.s3.us-east-1.amazonaws.com/db9a9d22-23d2-4070-86b6-d4b301d28c4b/myutils.zip

CHAPTER 5   GETTING STARTED WITH GENAI USING SAP BTP AND AMAZON BEDROCK

Step 6: Drag and drop the file to the root folder of your recently created Notebook.

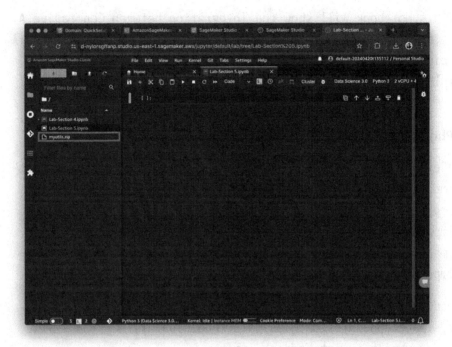

In the Amazon SageMaker Studio Notebook, click the **plus (+)** button to add cells and then type the code into the cells section by section. Click **Play** to see the outcome for each.

For the kernel, you can choose **Data Science 3.0 with Python 3** and **ml.t3.medium**.

253

Step 7: Copy and paste the following code into a cell; then click the **Play** button:

```
#This Notebook will show the integration between Generative AI
with SAP HANA Cloud
!pip install -q langchain boto3 awscli botocore
!pip install -q sqlalchemy-hana langchain_experimental hdbcli
```

Step 8: Copy and paste the following code into a cell; then click the **Play** button (changing the **bedrock region** if necessary):

```
#Setup boto3 client to access bedrock in a shared AWS Account
import json
import os
import sys
import boto3
import botocore

module_path = "/root/myutils.zip"
sys.path.append(os.path.abspath(module_path))
from myutils import bedrock, print_ww

# ---- ⚠ Un-comment and edit the below lines as needed for
your AWS setup ⚠ ----
os.environ["AWS_DEFAULT_REGION"] = "us-east-1"
# os.environ["AWS_PROFILE"] = "<YOUR_PROFILE>"
# os.environ["BEDROCK_ASSUME_ROLE"] = "arn:aws:iam::<SharedAWSA
ccount>:role/Crossaccountbedrock"   # E.g. "arn:aws:..."

boto3_bedrock = bedrock.get_bedrock_client(
    assumed_role=os.environ.get("BEDROCK_ASSUME_ROLE", None),
    region=os.environ.get("AWS_DEFAULT_REGION", None),
    runtime=True
)
```

CHAPTER 5 GETTING STARTED WITH GENAI USING SAP BTP AND AMAZON BEDROCK

Check the results to see if execution was successful.

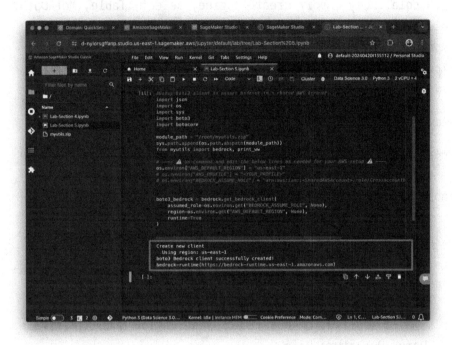

Step 9: Copy and paste the following code into a cell; then click the **Play** button:

```
import langchain
from langchain.sql_database import SQLDatabase
from langchain_experimental.sql import SQLDatabaseChain
from langchain.prompts.prompt import PromptTemplate
from langchain.llms.sagemaker_endpoint import LLMContentHandler
from langchain import SagemakerEndpoint
from langchain.llms.bedrock import Bedrock
from urllib.parse import quote
import sqlalchemy
from sqlalchemy import create_engine
```

255

```python
from sqlalchemy.orm import sessionmaker
from sqlalchemy import create_engine, select, Table, MetaData, Column, String
import hdbcli
import json
```

Step 10: Copy and paste the following code into a cell; then click the **Play** button:

```python
#Next step is to prepare the template for prompt and input to be used by the Generative AI
table_info_ar = [
    "Table Hotel has fields Name, Address, City, State, Zip code.",
    "Table Room has fields Free or Available, Price.",
    "Table Customer has fields Customer Number, title, first name, name, address, zip code.",
    "Table Reservation has fields Reservation Number,Arrival Date, Departure Date.",
    "Table Maintenance has fields Description, Date performed, Performed by."
]
table_info = "\n".join(table_info_ar)

_DEFAULT_TEMPLATE_ar = [
    "Given an input question, create a syntactically correct {dialect}  SQL query to run without comments, then provide answer to the question in english based on the result of SQL Query.",
    "Always use schema USER1.",
    "",
    "Use the following table:",
    "{table_info}",
    "",
```

CHAPTER 5   GETTING STARTED WITH GENAI USING SAP BTP AND AMAZON BEDROCK

```
    "Example:",
    "Human: How many Hotel are there ?",
    "SQLQuery: SELECT COUNT(*) AS num_hotels FROM HOTEL ",
    "SQLResult: [(3,)]",
    "Assistant: There are 3 hotels.",
    "",
    "Human: {input}",
    "",
    "Assistant: ",
]
_DEFAULT_TEMPLATE = "\n".join(_DEFAULT_TEMPLATE_ar)

PROMPT = PromptTemplate(
    input_variables=["input", "table_info", "dialect"],
    template=_DEFAULT_TEMPLATE
)
```

Step 11: Copy and paste the following code into a cell; then click the **Play** button:

```
#Next let's initiate the bedrock and the handler treatment of input and output
class ContentHandler(LLMContentHandler):
    content_type = "application/json"
    accepts = "application/json"

    def transform_input(self, prompt: str, model_kwargs: dict) -> bytes:
        input_str = json.dumps({"text_inputs": prompt, **model_kwargs})
        return input_str.encode('utf-8')

    def transform_output(self, output: bytes) -> str:
        response_json = json.loads(output.read().decode("utf-8"))
        return response_json["generated_texts"][0]
```

CHAPTER 5   GETTING STARTED WITH GENAI USING SAP BTP AND AMAZON BEDROCK

```
content_handler = ContentHandler()

model_parameter = {"temperature": 0, "max_tokens_to_
sample": 600}

#Make sure your account has access to anthropic claude access.
This can be enabled from Bedrock console. Access is auto
approved.
llm = Bedrock(model_id="anthropic.claude-v2", model_
kwargs=model_parameter, client=boto3_bedrock)
```

If you encounter a warning in red text like the one shown here, just ignore it:

Step 12: Copy and paste the following code into a cell; then click the **Play** button (remembering to change the user, password, and SQL endpoint):

258

CHAPTER 5   GETTING STARTED WITH GENAI USING SAP BTP AND AMAZON BEDROCK

```
#Let's connect to the SAP HANA Database and then execute
langchain SQL Database Chain to query from the Generative AI
db = SQLDatabase.from_uri("hana://<user>:<password>@<changesqle
ndpoint>.hana.trial-us10.hanacloud.ondemand.com:443")
db_chain = SQLDatabaseChain.from_llm(llm=llm, db=db,
prompt=PROMPT, verbose=True, use_query_checker=True, top_k=10 )
```

Example: "USER1, Password1, 73098405-c4b1-437c-89cf-3248286c20fd"

Step 13: Copy and paste the following code into a cell; then click the **Play** button:

```
#Execute the first query, this is a simple English to text SQL
db_chain.invoke("How many HOTEL are there ?")
```

Now you will see an error related to access being denied to the LLM model:

CHAPTER 5 GETTING STARTED WITH GENAI USING SAP BTP AND AMAZON BEDROCK

Step 14: Go to the Amazon Bedrock **Model access** page.

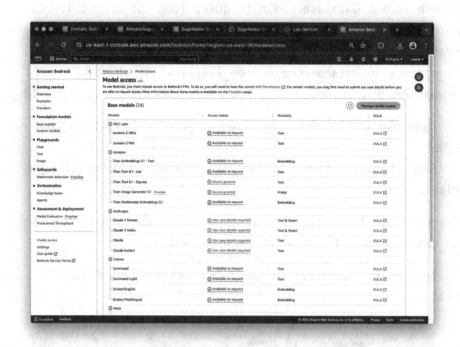

CHAPTER 5   GETTING STARTED WITH GENAI USING SAP BTP AND AMAZON BEDROCK

Step 15: Click **Manage model access**.

Step 16: Click **Submit use case**. The following image is an example:

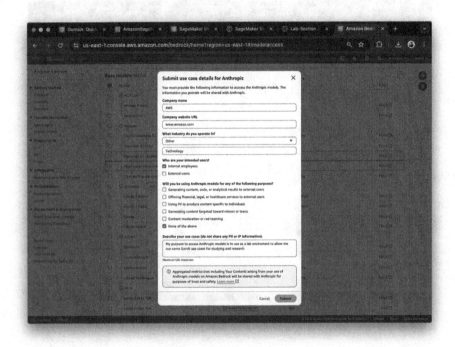

CHAPTER 5   GETTING STARTED WITH GENAI USING SAP BTP AND AMAZON BEDROCK

Step 17: Select **Claude3 Sonnet**, **Claude**, and **Claude Instant**.

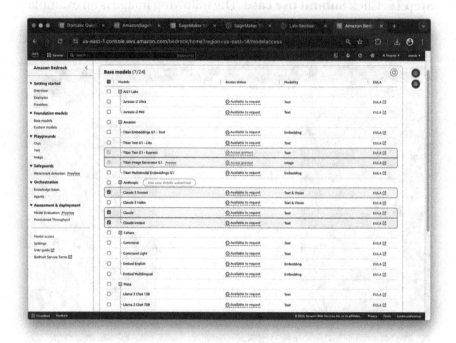

CHAPTER 5 GETTING STARTED WITH GENAI USING SAP BTP AND AMAZON BEDROCK

Step 18: Scroll down the page and click the **Save changes** button.

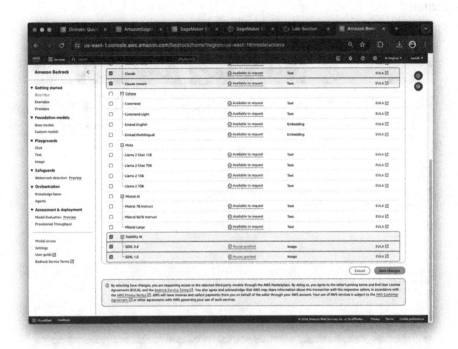

CHAPTER 5   GETTING STARTED WITH GENAI USING SAP BTP AND AMAZON BEDROCK

Now you should see the models enabled, indicated by "Access Granted."

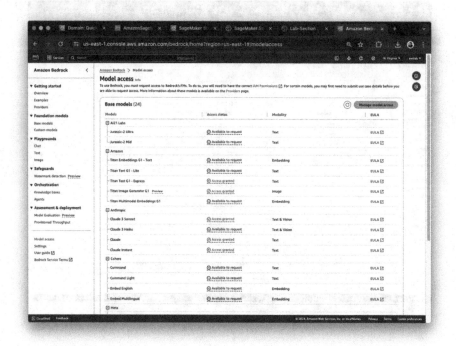

CHAPTER 5   GETTING STARTED WITH GENAI USING SAP BTP AND AMAZON BEDROCK

Step 19: Go to the notebook and click **Play** in the cell.
You can now see the results:

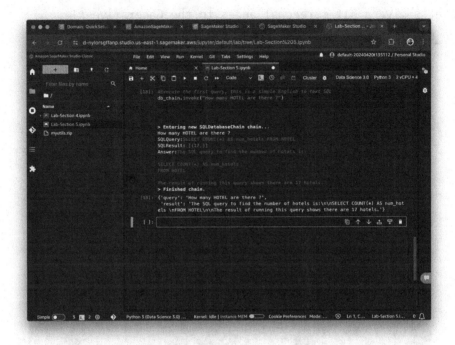

CHAPTER 5   GETTING STARTED WITH GENAI USING SAP BTP AND AMAZON BEDROCK

Step 20: Copy and paste the following code into the cell; then click the **Play** button:

```
#Let's give a bit more challenges to the Generative AI
db_chain.invoke("How many HOTEL are there in Seattle ?")
```

Now you can see the results:

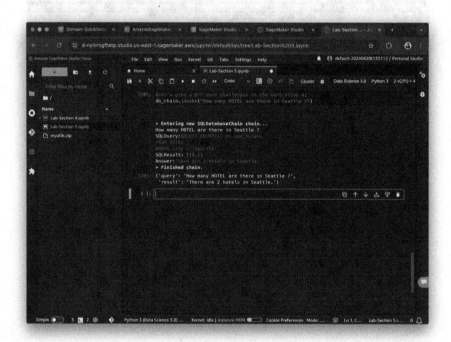

CHAPTER 5   GETTING STARTED WITH GENAI USING SAP BTP AND AMAZON BEDROCK

Step 21: Copy and paste the following code into a cell; then click the **Play** button:

```
#And what happened if the information is stored in
another table
db_chain.invoke("How many free single and double rooms are
there in Congress Hotel ?")
```

Now you can see the results:

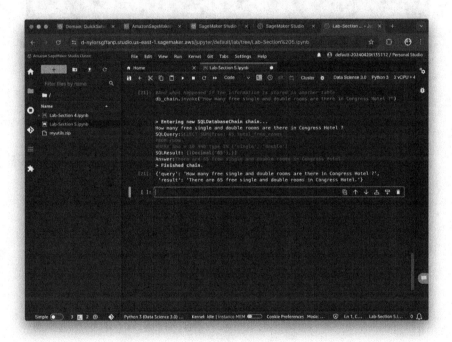

Congratulations! You have successfully concluded this section's exercise.

CHAPTER 5   GETTING STARTED WITH GENAI USING SAP BTP AND AMAZON BEDROCK

Experiment with Langchain SQL Agent

Step 1: Choose **File ➤ New ➤ Notebook.**

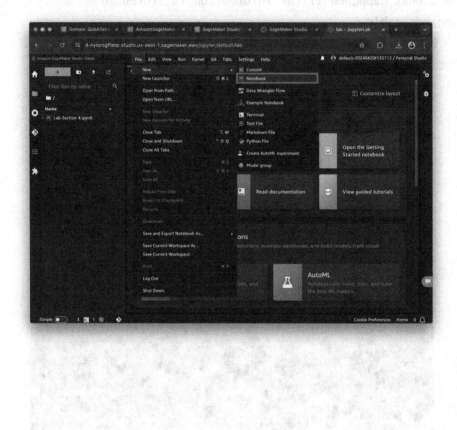

CHAPTER 5 GETTING STARTED WITH GENAI USING SAP BTP AND AMAZON BEDROCK

Step 2: Click the **Select** button.

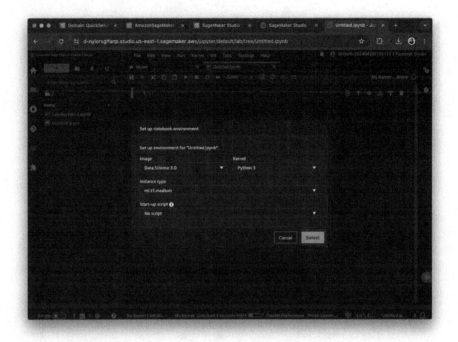

CHAPTER 5   GETTING STARTED WITH GENAI USING SAP BTP AND AMAZON BEDROCK

Step 3: Rename the notebook.

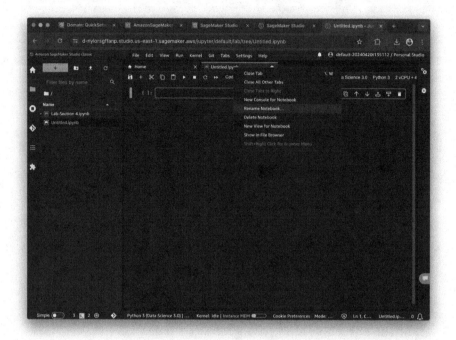

CHAPTER 5   GETTING STARTED WITH GENAI USING SAP BTP AND AMAZON BEDROCK

Step 4: Click the **Rename** button.

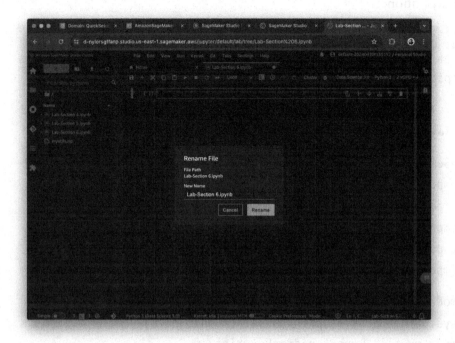

In the Amazon SageMaker Studio Notebook, click the **plus (+)** button to add cells; then you can type the code into the cells section by section and click **Play** to see the outcome for each.

For the kernel, you can choose **Data Science 3.0** with **Python 3** and **ml.t3.medium**.

Step 5: Copy and paste the following code into a cell; then click the **Play** button:

```
#This Notebook will show the integration between Generative AI with SAP HANA Cloud

!pip install -q langchain hdbcli
```

Step 6: Copy and paste the following code into a cell, then click the **Play** button:

```
#Setup boto3 client to access bedrock in a shared AWS Account
import json
import os
import sys
import boto3
import botocore

module_path = "/root/myutils.zip"
sys.path.append(os.path.abspath(module_path))
from myutils import bedrock, print_ww

# ---- ⚠ Un-comment and edit the below lines as needed for your AWS setup ⚠ ----
os.environ["AWS_DEFAULT_REGION"] = "us-east-1"
# os.environ["AWS_PROFILE"] = "<YOUR_PROFILE>"
# os.environ["BEDROCK_ASSUME_ROLE"] = "arn:aws:iam::<SharedAWSAccount>:role/Crossaccountbedrock"   # E.g. "arn:aws:..."

boto3_bedrock = bedrock.get_bedrock_client(
    assumed_role=os.environ.get("BEDROCK_ASSUME_ROLE", None),
    region=os.environ.get("AWS_DEFAULT_REGION", None),
    runtime=True)
```

CHAPTER 5 GETTING STARTED WITH GENAI USING SAP BTP AND AMAZON BEDROCK

Look at the results to determine if the execution was successful.

Step 7: Copy and paste the following code into a cell; then click the **Play** button:

```
from langchain.agents import create_sql_agent
from langchain.agents.agent_toolkits import SQLDatabaseToolkit
from langchain.agents.agent_types import AgentType
from langchain.sql_database import SQLDatabase
from langchain.llms.bedrock import Bedrock
import json
```

273

CHAPTER 5   GETTING STARTED WITH GENAI USING SAP BTP AND AMAZON BEDROCK

Step 8: Copy and paste the following code into a cell; then click the **Play** button:

model_parameter = {"temperature": 0, "max_tokens_to_sample": 600}
langchain_llm = Bedrock(model_id="anthropic.claude-v2:1", model_kwargs=model_parameter, client=boto3_bedrock)
ignore

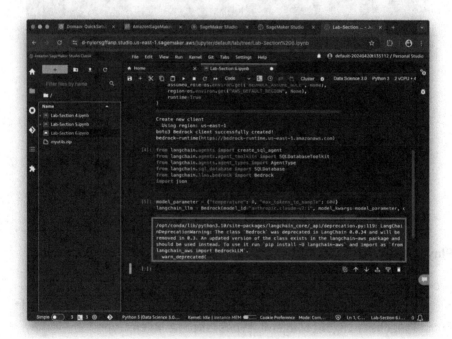

CHAPTER 5 GETTING STARTED WITH GENAI USING SAP BTP AND AMAZON BEDROCK

Step 9: Copy and paste the following code into a cell; then click the **Play** button:

```
#Let's connect to the SAP HANA Database and then execute Agent with SQL Database Toolkit
db = SQLDatabase.from_uri("hana://<user>:<password>@<changesqlendpoint>.hana.trial-us10.hanacloud.ondemand.com:443")
toolkit = SQLDatabaseToolkit(db=db, llm=langchain_llm)
```

Remember to change the user, password, and SQL endpoint.

Step 10: Copy and paste the following code into a cell; then click the **Play** button:

```
agent_executor = create_sql_agent(
    llm=langchain_llm,
    toolkit=toolkit,
    verbose=True,
    agent_type=AgentType.ZERO_SHOT_REACT_DESCRIPTION,
    handle_parsing_errors=True
)
```

Step 11: Copy and paste the following code into a cell; then click the **Play** button:

```
#Let's give a bit more challenges to the Generative AI
res = agent_executor({"input":"How many Hotels are there in Seattle ?"})
print(res['output'])
```

CHAPTER 5 GETTING STARTED WITH GENAI USING SAP BTP AND AMAZON BEDROCK

In the results, just ignore the warning message.

CHAPTER 5   GETTING STARTED WITH GENAI USING SAP BTP AND AMAZON BEDROCK

Step 12: Scroll down the page to see more results.

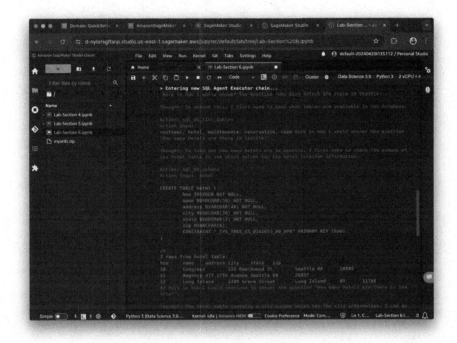

CHAPTER 5   GETTING STARTED WITH GENAI USING SAP BTP AND AMAZON BEDROCK

Scroll down a little bit more to analyze the final result.

Step 13: Copy and paste the following code into a cell; then click the **Play** button:

```
#And what happened if the information is stored in another table
res = agent_executor({"input":"How many total free rooms are there in Congress Hotel ?"})
print(res['output'])
```

CHAPTER 5 GETTING STARTED WITH GENAI USING SAP BTP AND AMAZON BEDROCK

Step 14: Scroll down the page to see more results.

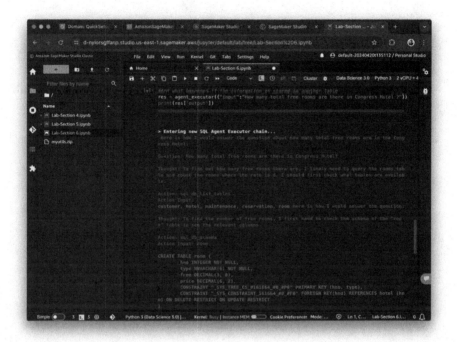

CHAPTER 5   GETTING STARTED WITH GENAI USING SAP BTP AND AMAZON BEDROCK

Scroll down a bit more to analyze the final result.

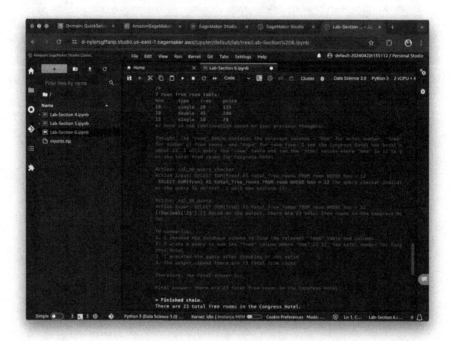

Step 15: Copy and paste the following code into a cell; then click the **Play** button:

```
#How about a more challenging and complex query ?
res = agent_executor({"input":"What are the maintenances done at Congress Hotel ?"})
print(res['output'])
```

CHAPTER 5 GETTING STARTED WITH GENAI USING SAP BTP AND AMAZON BEDROCK

Step 16: Scroll down the page to see more results.

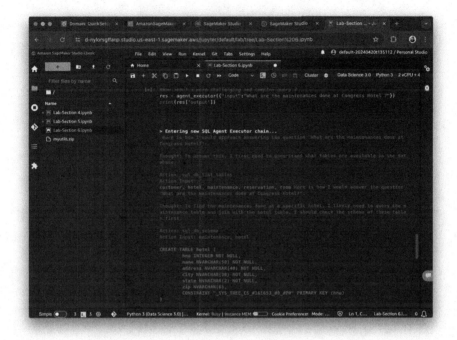

CHAPTER 5   GETTING STARTED WITH GENAI USING SAP BTP AND AMAZON BEDROCK

Once again, scroll down a bit to analyze the final result.

CHAPTER 5   GETTING STARTED WITH GENAI USING SAP BTP AND AMAZON BEDROCK

Finally, you get a better response at the bottom of the following image:

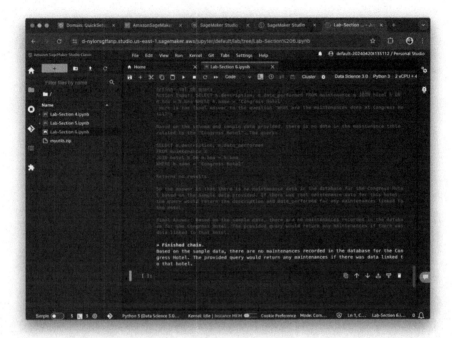

Congratulations! You have successfully concluded all of the exercises for this section.

We've worked with aspects related to GenAI and experimented with how LLMs can support you in a structured result from a given question. In the process, we noted the importance of employing **observation** and **thought** before providing **final answers from the results obtained**.

In this chapter, you learned how to get started with generative AI use cases utilizing Amazon SageMaker Notebooks. Doing so has shown you how to take advantage of a GenAI LLM and improve the user experience when exploring business data stored in SAP HANA Cloud. You also had the chance to use Amazon Bedrock, the Python language, and the Langchain SQL agent.

In the next chapter, we will continue our journey with a few more scenarios.

CHAPTER 6

Building GenAI with SAP, Lambda, and Amazon Bedrock

You have seen how Amazon SageMaker Notebook integrates a GenAI LLM to improve the user's experience by exploring business data stored in SAP HANA Cloud. In this chapter, we will explore other scenarios using Amazon Bedrock, the Python language, Langchain, PandasAI, and SQLAlchemy.

The architecture used in this book allows you to leverage several generative AI models, such as the Amazon Titan, FLAN T5, and Claude. To leverage other models, you just need to deploy the related endpoints and access the service through API calls from Notebook. Note that the performance of the solution really depends on the deployed pretrained model's quality and performance.

CHAPTER 6   BUILDING GENAI WITH SAP, LAMBDA, AND AMAZON BEDROCK

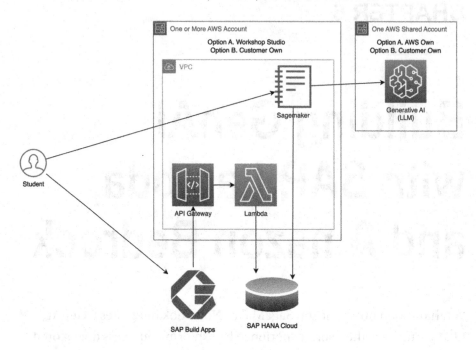

Based on the architecture in the previous image, a common workflow would consist of the following steps:

1. The user types a prompt in English such as "Which hotels are the most profitable?"

2. The Python code will then pass the prompt to a generative AI model.

3. Generative AI will generate the required SQL Query statement.

4. The SQL Query will be executed in SAP HANA Cloud through its secured protocol and return a result.

5. Generative AI will provide an answer based on the result from SAP HANA Cloud

6. Generative AI can also be exposed to an API to be accessed from other systems (e.g., SAP Build Apps).

CHAPTER 6   BUILDING GENAI WITH SAP, LAMBDA, AND AMAZON BEDROCK

Analyze a SAP Report Using Generative AI

In this exercise, we will put into practice some possibilities of AI using SAP reports and Amazon Bedrock with the following focus:

- Unlocking a generative AI API to analyze SAP reports
- Taking advantage of GenAI for the summarization of PDF files
- Leveraging the Claude LLM to simulate a financial analyst

Step 1: Choose **File** ➤ **new** ➤ **Notebook.**

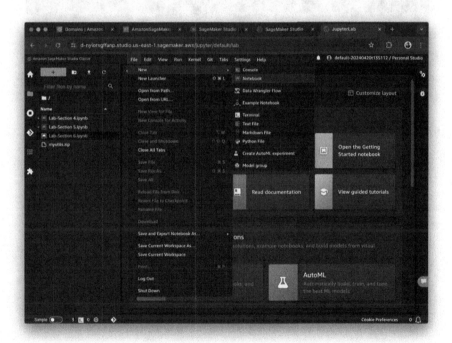

CHAPTER 6 BUILDING GENAI WITH SAP, LAMBDA, AND AMAZON BEDROCK

Step 2: Click the **Select** button.

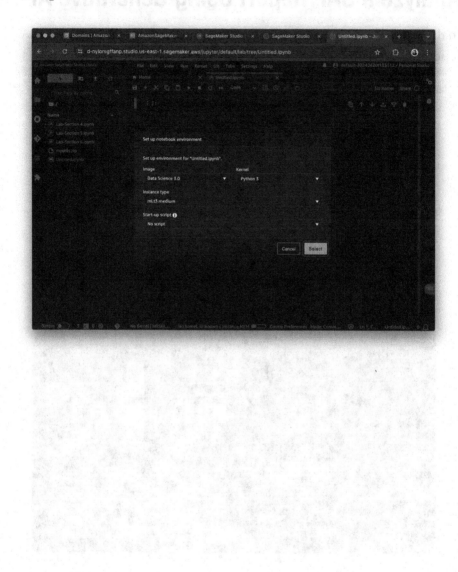

CHAPTER 6   BUILDING GENAI WITH SAP, LAMBDA, AND AMAZON BEDROCK

Step 3: Rename the notebook.

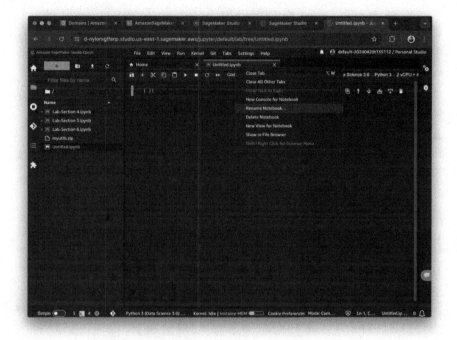

289

Step 4: Click the **Rename** button.

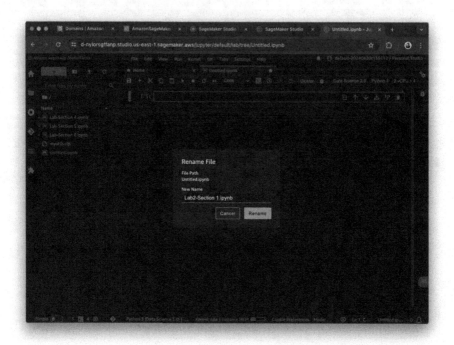

Step 5: Download the file `myutils.zip` to your computer.
https://ws-assets-prod-iad-r-iad-ed304a55c2ca1aee.s3.us-east-1.amazonaws.com/db9a9d22-23d2-4070-86b6-d4b301d28c4b/myutils.zip

CHAPTER 6 BUILDING GENAI WITH SAP, LAMBDA, AND AMAZON BEDROCK

Step 6: Drag and drop the file to the root folder of the recently created Notebook.

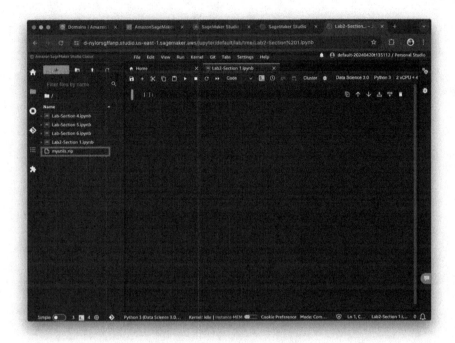

Step 7: Download RFSSLD00.zip to your computer.

https://ws-assets-prod-iad-r-iad-ed304a55c2ca1aee.s3.us-east-1.amazonaws.com/db9a9d22-23d2-4070-86b6-d4b301d28c4b/RFSSLD00.pdf

CHAPTER 6   BUILDING GENAI WITH SAP, LAMBDA, AND AMAZON BEDROCK

Step 8: Drag and drop the file to the root folder of the recently created notebook.

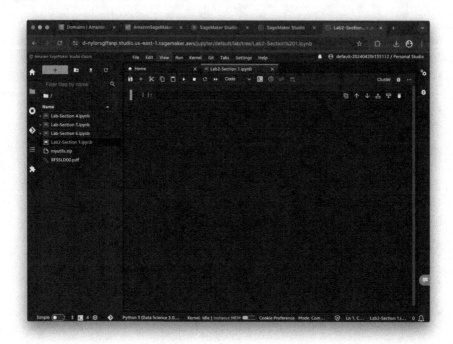

CHAPTER 6 BUILDING GENAI WITH SAP, LAMBDA, AND AMAZON BEDROCK

You can preview the content of the PDF file as follows:

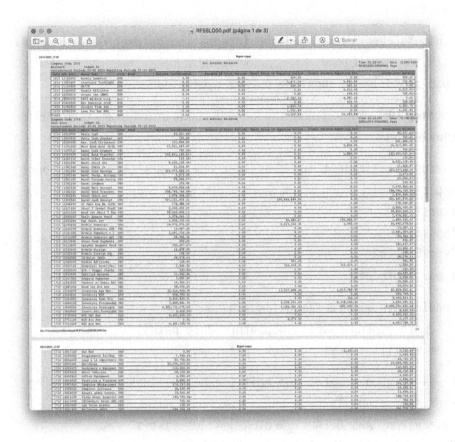

In the Amazon SageMaker Studio Notebook, click the **plus (+)** button (as shown in the following image) to add cells, type the code into the cells section by section, and then click **Play** to see the outcome for each.

For the kernel, you can choose **Data Science 3.0 with Python 3** and **ml.t3.medium**.

Chapter 6   Building GenAI with SAP, Lambda, and Amazon Bedrock

Step 9: Copy and paste the following code into a cell; then click the **Play** button:

```
#This Notebook will show the integration between Generative AI with SAP HANA Cloud
!pip install -q --upgrade pyPDF2 pycryptodome
```

Step 10: Copy and paste the following code into the cell; then click the **Play** button:

```
#Setup boto3 client to access bedrock in a shared AWS Account
import json
import os
import sys
import boto3
import botocore

module_path = "/root/myutils.zip"
sys.path.append(os.path.abspath(module_path))
from myutils import bedrock, print_ww

# ---- ⚠ Un-comment and edit the below lines as needed for your AWS setup ⚠ ----
os.environ["AWS_DEFAULT_REGION"] = "us-east-1"
# os.environ["AWS_PROFILE"] = "<YOUR_PROFILE>"
# os.environ["BEDROCK_ASSUME_ROLE"] = "arn:aws:iam::<SharedAWSAccount>:role/Crossaccountbedrock"   # E.g. "arn:aws:..."

boto3_bedrock = bedrock.get_bedrock_client(
    assumed_role=os.environ.get("BEDROCK_ASSUME_ROLE", None),
    region=os.environ.get("AWS_DEFAULT_REGION", None),
    runtime=True
)
```

CHAPTER 6   BUILDING GENAI WITH SAP, LAMBDA, AND AMAZON BEDROCK

Observe the results:

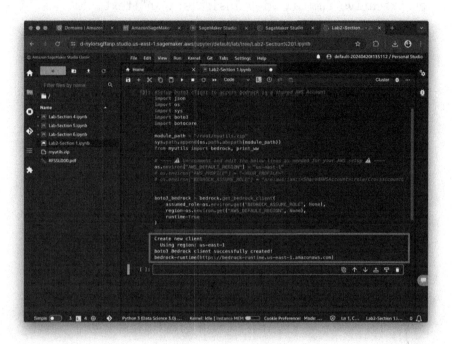

Step 11: Copy and paste the following code into a cell; then click the **Play button:**

```
import os
from PyPDF2 import PdfReader

def summarize_pdf(path: str) -> str:
    reader = PdfReader(path)
    text = "\n".join([page.extract_text() for page in reader.pages])

    System = "You are a great Finance Analyst, which can analyze the health of a company based on a Financial Report."
```

```python
User = "Analyze the result of the Report below and provide 
an Executive Summary with Insights.\n\n"+text

body = json.dumps({
  "max_tokens": 2000,
  "temperature": 0,
  "system": System,
  "messages": [
      {
          "role": "user",
          "content": [
              {
                  "type": "text",
                  "text": User
              }
          ]
      }
  ],
  "anthropic_version": "bedrock-2023-05-31"
})
response = boto3_bedrock.invoke_model(body=body, 
modelId="anthropic.claude-3-sonnet-20240229-v1:0")
response_body = json.loads(response.get("body").read())
result = response_body.get("content")
return result
```

CHAPTER 6 BUILDING GENAI WITH SAP, LAMBDA, AND AMAZON BEDROCK

Step 12: Copy and paste the following code into a cell; then click the **Play** button:

```
analysis=summarize_pdf("RFSSLD00.pdf")
print(analysis[0]['text'])
```

Observe the results:

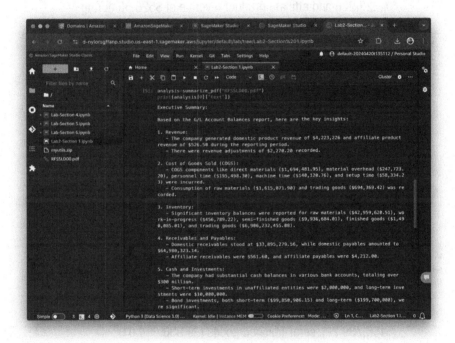

Observe the executive summary created by the generative AI with Claude's LLM using an SAP General Ledger balance report. This can truly improve productivity for a financial manager.

297

CHAPTER 6   BUILDING GENAI WITH SAP, LAMBDA, AND AMAZON BEDROCK

EXECUTIVE SUMMARY:

Based on the G/L Account Balances report, here are the key insights:

- Revenue:

    The company generated domestic product revenue of $4,223,226 and affiliate product revenue of $526.50 during the reporting period.

    There were revenue adjustments of $2,270.20 recorded.

- Cost of Goods Sold (COGS):

    COGS components like direct materials ($1,694,481.95), material overhead ($247,723.20), personnel time ($195,498.30), machine time ($140,120.76), and setup time ($58,334.23) were incurred.

    Consumption of raw materials ($1,615,071.90) and trading goods ($694,369.42) was recorded.

- Inventory:

    Significant inventory balances were reported for raw materials ($42,959,620.51), work-in-progress ($456,789.22), semi-finished goods ($9,936,684.01), finished goods ($1,490,085.01), and trading goods ($6,906,232,455.08).

- Receivables and Payables:

    Domestic receivables stood at $33,895,279.56, while domestic payables amounted to $64,980,323.14.

    Affiliate receivables were $561.60, and affiliate payables were $4,212.00.

- Cash and Investments:

    The company had substantial cash balances in various bank accounts, totaling more than $300 million.

    Short-term investments in unaffiliated entities were $2,000,000, and long-term investments were $10,000,000.

    Bond investments, both short-term ($99,850,906.15) and long-term ($199,700,000), were significant.

- Debt:

    Short-term bond payables were $600,000,000, while long-term bond payables were $100,000,000.

- Other Expenses:

    Bad debt expenses of $724,659.31 was recorded.

    Interest income from unaffiliated sources ($23,233.28) and bonds ($873,113.14) was earned.

    Interest expense on unaffiliated debt ($53,083.34) and bonds ($33,972.18) was incurred.

- Fixed Assets and Depreciation:

    The company had significant fixed assets, including buildings, machinery, vehicles, and computer equipment.

    Corresponding accumulated depreciation balances were reported for these assets.

Overall, the report provides a comprehensive view of the company's financial position, including its revenue streams, cost structure, inventory levels, receivables and payables, cash and investment holdings, debt obligations, and fixed asset base.

Congratulations! You have successfully concluded this section's exercise.

Integrate Generative AI with SAP HANA and PandasAI

By the end of this lab, you will be able to do the following:

- Explain the capabilities of PandasAI for simplifying data analysis workflows in Python
- Load data from SAP HANA views (which are the main components for reporting) into a Pandas data frame in SageMaker Studio Notebook
- Combine PandasAI and Claude to perform interactive data analysis of SAP HANA Cloud data
- Use PandasAI to generate visualizations, statistics, and insights from SAP HANA data
- Automate data analysis workflows for SAP HANA data using PandasAI and Claude in SageMaker Studio Notebook
- Troubleshoot errors when connecting SageMaker Studio Notebook to SAP HANA Cloud and utilizing PandasAI and Claude

CHAPTER 6 BUILDING GENAI WITH SAP, LAMBDA, AND AMAZON BEDROCK

Step 1: Choose **File ➤ new Notebook.**

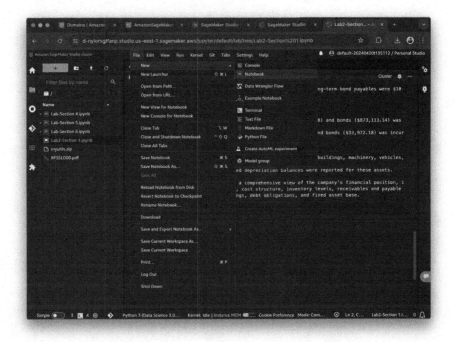

CHAPTER 6   BUILDING GENAI WITH SAP, LAMBDA, AND AMAZON BEDROCK

Step 2: Click the **Select** button.

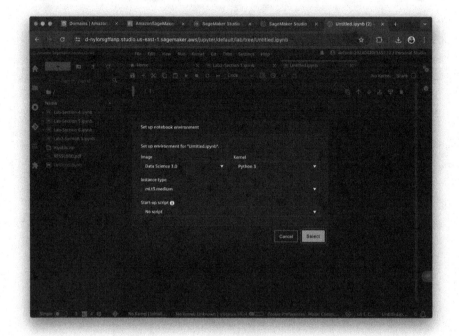

CHAPTER 6   BUILDING GENAI WITH SAP, LAMBDA, AND AMAZON BEDROCK

Step 3: Rename the notebook.

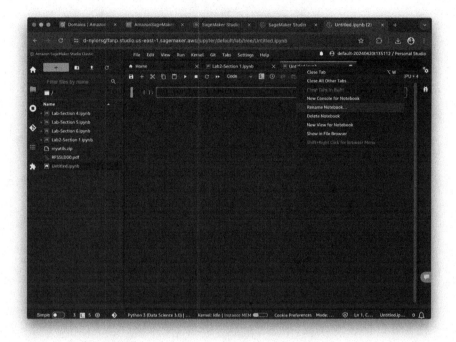

CHAPTER 6   BUILDING GENAI WITH SAP, LAMBDA, AND AMAZON BEDROCK

Step 4: Click the **Rename** button.

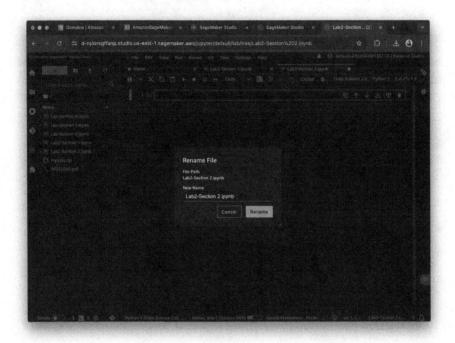

Step 5: Download `myutils.zip` to your computer.
https://ws-assets-prod-iad-r-iad-ed304a55c2ca1aee.s3.us-east-1.amazonaws.com/db9a9d22-23d2-4070-86b6-d4b301d28c4b/RFSSLD00.pdf

CHAPTER 6   BUILDING GENAI WITH SAP, LAMBDA, AND AMAZON BEDROCK

Step 6: Drag and drop the file to the root folder of your recently created notebook.

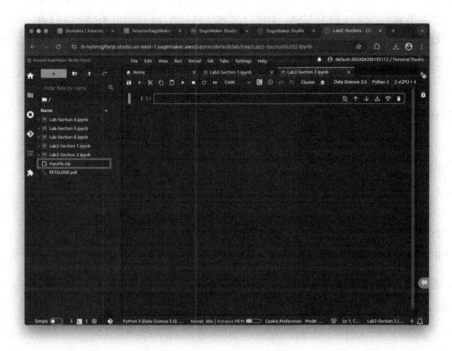

In the Amazon SageMaker Studio Notebook, click the **plus (+)** button to add cells; then you can type the code into the cells section by section and click **Play** to see the outcome for each.

For the kernel, you can choose **Data Science 3.0 with Python 3** and **ml.t3.medium**.

Step 7: Copy and paste the following code into a cell; then click the **Play** button:

```
#This Notebook will show the integration between Generative AI
with SAP HANA Cloud and PandasAI
!pip install -q hdbcli pandas pandasai sagemaker polars
```

Step 8: Copy and paste the following code into a cell; then click the **Play** button:

```
#Setup boto3 client to access bedrock in a shared AWS Account
import json
import os
import sys
import boto3
import botocore

module_path = "/root/myutils.zip"
sys.path.append(os.path.abspath(module_path))
from myutils import bedrock, print_ww

# ---- ⚠ Un-comment and edit the below lines as needed for
your AWS setup ⚠ ----
os.environ["AWS_DEFAULT_REGION"] = "us-east-1"
# os.environ["AWS_PROFILE"] = "<YOUR_PROFILE>"
# os.environ["BEDROCK_ASSUME_ROLE"] = "arn:aws:iam::<SharedAWSA
ccount>:role/Crossaccountbedrock"   # E.g. "arn:aws:..."

boto3_bedrock = bedrock.get_bedrock_client(
    assumed_role=os.environ.get("BEDROCK_ASSUME_ROLE", None),
    region=os.environ.get("AWS_DEFAULT_REGION", None),
    runtime=True
)
```

CHAPTER 6 BUILDING GENAI WITH SAP, LAMBDA, AND AMAZON BEDROCK

Observe the results:

Step 9: Copy and paste the following code into a cell; then click the **Play** button:

```
#Import your dependencies
from hdbcli import dbapi
import pandas as pd
from pandasai import SmartDataframe
from pandasai.llm import BedrockClaude
```

Step 10: Copy and paste the following code into a cell; then click the **Play** button (remembering to change the user, password, and SQL endpoint):

```
#Initialize your connection
conn = dbapi.connect(
    address='<changesqlendpoint.hana.trial-us10.hanacloud.
    ondemand.com>',
```

307

CHAPTER 6   BUILDING GENAI WITH SAP, LAMBDA, AND AMAZON BEDROCK

```
    port='443',
    user='<ChangeUser>',
    password='<ChangePassword>',
    encrypt=True,
    sslValidateCertificate=True
)
#If no errors, print connected
print('connected\n')
```

Example: "USER1, Password1,

73098405-c4b1-437c-89cf-3248286c20fd"

If you see an error as highlighted in the following image, check if the SAP HANA Cloud instance is down.

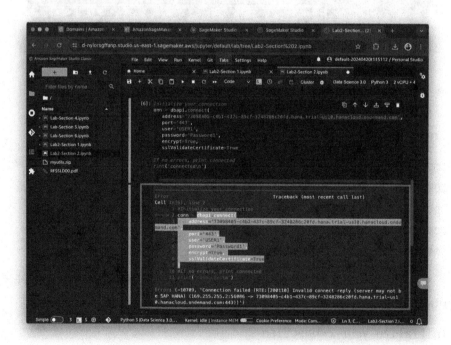

308

CHAPTER 6 BUILDING GENAI WITH SAP, LAMBDA, AND AMAZON BEDROCK

Step 11: To do so, go to the SAP BTP Cockpit and click **Manage SAP HANA Cloud**.

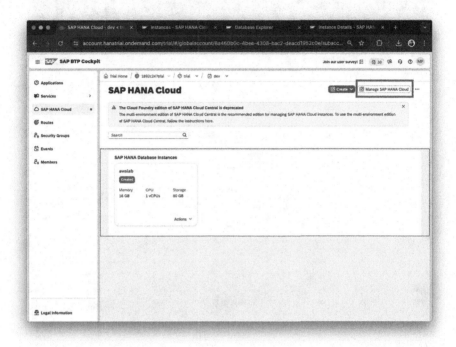

CHAPTER 6   BUILDING GENAI WITH SAP, LAMBDA, AND AMAZON BEDROCK

Step 12: Start the SAP HANA Cloud instance.

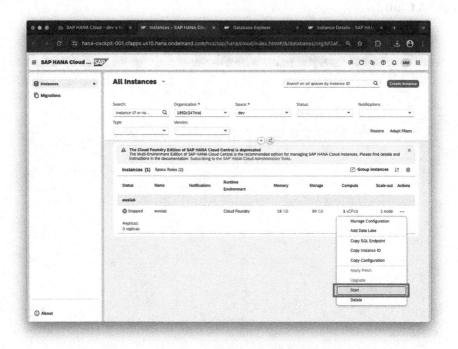

CHAPTER 6 BUILDING GENAI WITH SAP, LAMBDA, AND AMAZON BEDROCK

You can see that the instance status has changed to Starting.

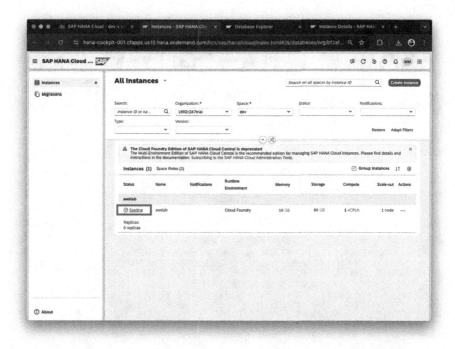

CHAPTER 6   BUILDING GENAI WITH SAP, LAMBDA, AND AMAZON BEDROCK

After a few minutes, the SAP HANA Cloud instance will be up and running.

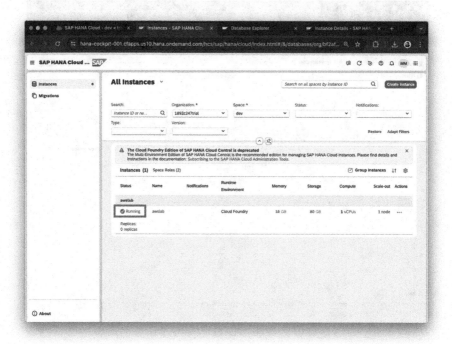

CHAPTER 6   BUILDING GENAI WITH SAP, LAMBDA, AND AMAZON BEDROCK

Step 13: Now go back to Amazon SageMaker Notebook and click **Play** for the last cell code. You should see that the connection will be successfully established.

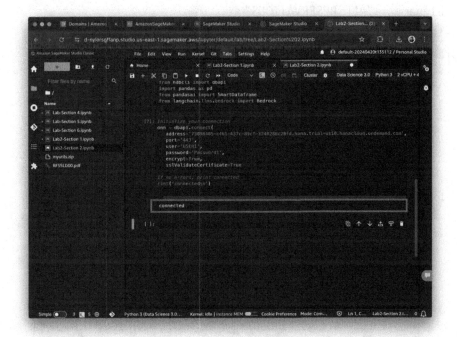

CHAPTER 6   BUILDING GENAI WITH SAP, LAMBDA, AND AMAZON BEDROCK

Step 14: Copy and paste the following code into a cell; then click the **Play** button:

```
schema = "USER1"
tablename = "ALL_RESERVATIONS"
data=pd.read_sql(f'select * from {schema}.{tablename}',conn)
print(data)
```

Observe the results:

CHAPTER 6   BUILDING GENAI WITH SAP, LAMBDA, AND AMAZON BEDROCK

Step 15: Copy and paste the following code into a cell; then click the **Play** button:

```
# Instantiate a LLM
llm = BedrockClaude(boto3_bedrock)
df = SmartDataframe(data, config={"llm": llm})
```

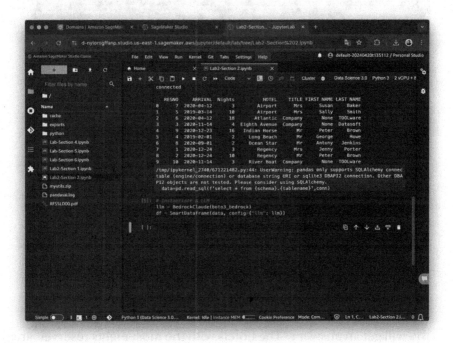

CHAPTER 6   BUILDING GENAI WITH SAP, LAMBDA, AND AMAZON BEDROCK

Step 16: Copy and paste the following code into a cell; then click the **Play** button:

```
df.chat("Which Hotel has the most number of nights reserved?")
```

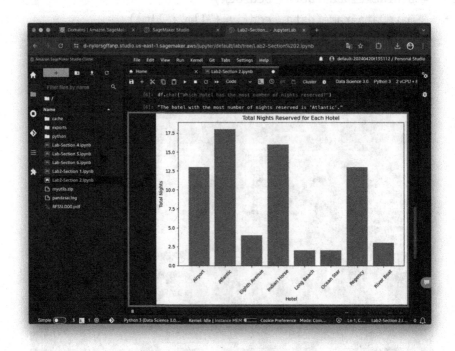

You will get an answer : "The hotels with the most nights reserved are: Atlantic."

CHAPTER 6 BUILDING GENAI WITH SAP, LAMBDA, AND AMAZON BEDROCK

Step 17: Copy and paste the following code into a cell; then click the **Play** button:

df.chat("List the hotel name sorted by its total number of nights reserved")

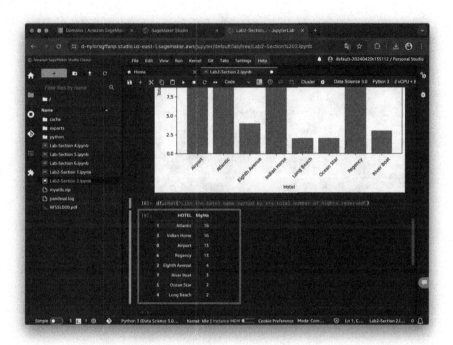

CHAPTER 6   BUILDING GENAI WITH SAP, LAMBDA, AND AMAZON BEDROCK

Step 18: Copy and paste the following code into a cell; then click the **Play** button:

df.chat("What is the sum of nights for all the hotels ?")

You should get an answer of **71**.

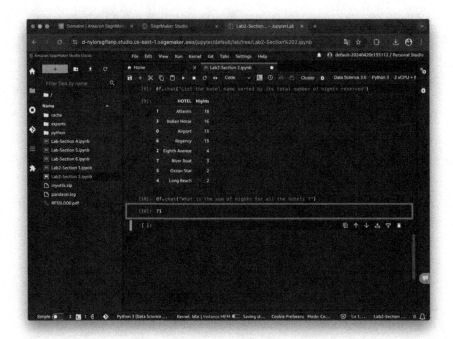

CHAPTER 6   BUILDING GENAI WITH SAP, LAMBDA, AND AMAZON BEDROCK

Step 19: Copy and paste the following code into a cell; then click the **Play button:**

```
df.chat("Provide Full Analysis of this data for a Hotel Manager.")
```

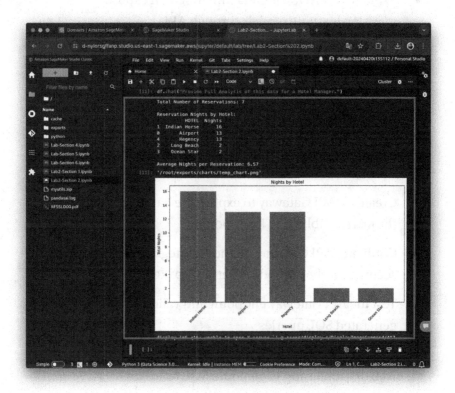

Congratulations! You have successfully concluded this section.

CHAPTER 6 BUILDING GENAI WITH SAP, LAMBDA, AND AMAZON BEDROCK

Building Generative AI Using Lambda and API Gateway

By the end of this lab, you will be able to do the following:

- Explain how AWS Lambda functions can expose machine learning models through APIs

- Package the model code and dependencies into a Lambda deployment package (layers and functions)

- Write a Lambda function handler to accept text prompts and return SQL queries

- Deploy the Lambda function within AWS serverless infrastructure

- Create an API Gateway to expose the Lambda function through a public HTTP endpoint

- Configure API Gateway methods, integration, and security for the Lambda function (apikey)

- Test the API Gateway endpoint and validate the generated SQL queries

- Call the API endpoint from a Postman-simulating a web or mobile app to generate SQL from text

- Troubleshoot issues with the Lambda function, the API Gateway, and the text-to-SQL model

Choose **File ➤ new ➤ Notebook** and then start the terminal by clicking the **$ (terminal)** button on the SageMaker Studio Notebook. Execute the scripts as follows:

CHAPTER 6   BUILDING GENAI WITH SAP, LAMBDA, AND AMAZON BEDROCK

Step 1: Choose **File** ➤ **new** ➤ **Notebook**.

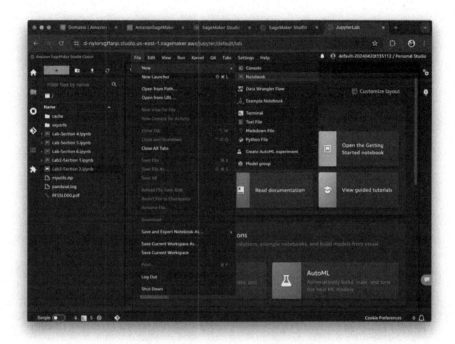

For the kernel, you can choose **Data Science 3.0 with Python 3** and **ml.m5.large**.

CHAPTER 6   BUILDING GENAI WITH SAP, LAMBDA, AND AMAZON BEDROCK

Step 2: Click the **Select** button.

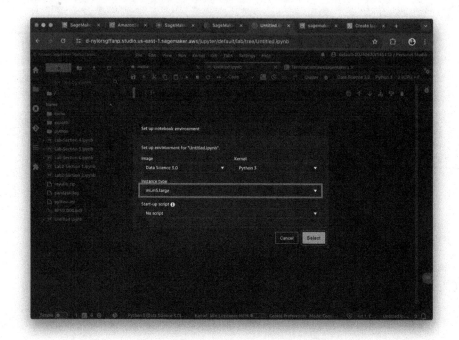

CHAPTER 6   BUILDING GENAI WITH SAP, LAMBDA, AND AMAZON BEDROCK

Step 3: Rename the notebook.

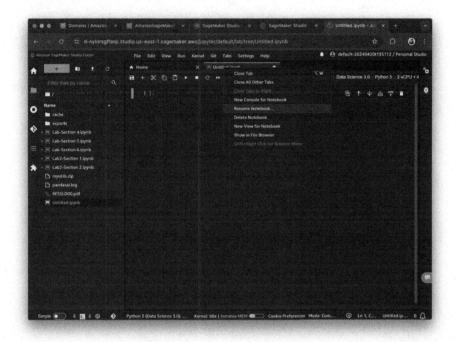

CHAPTER 6   BUILDING GENAI WITH SAP, LAMBDA, AND AMAZON BEDROCK

Step 4: Click the **Rename** button.

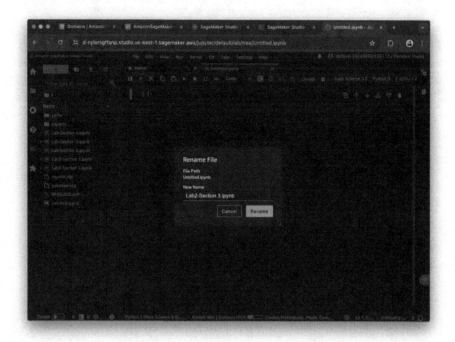

Step 5: Download `myutils.zip` to your computer.

Please change *AWSAccountNumber* to your own AWS account number and create the `lambda_layer` folder in the bucket `s3://sagemaker-us-east-1-AWSAccountNumber/`.

CHAPTER 6   BUILDING GENAI WITH SAP, LAMBDA, AND AMAZON BEDROCK

Step 6: Go to Amazon S3.

https://us-east-1.console.aws.amazon.com/s3/home?region=us-east-1#

You should see this panel:

CHAPTER 6   BUILDING GENAI WITH SAP, LAMBDA, AND AMAZON BEDROCK

Step 7: Click **Amazon S3**.

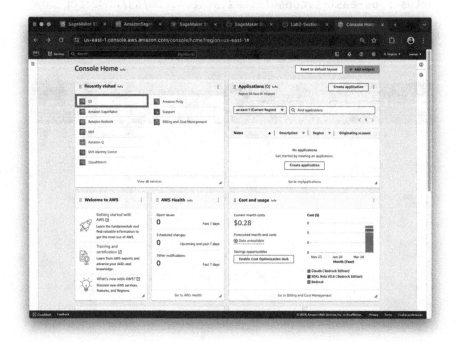

CHAPTER 6   BUILDING GENAI WITH SAP, LAMBDA, AND AMAZON BEDROCK

Step 8: Click the **Create bucket** button.

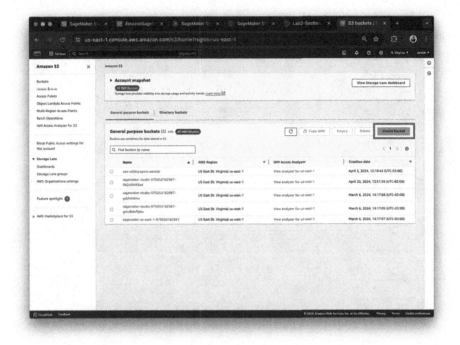

CHAPTER 6 BUILDING GENAI WITH SAP, LAMBDA, AND AMAZON BEDROCK

Step 9: Name the bucket.

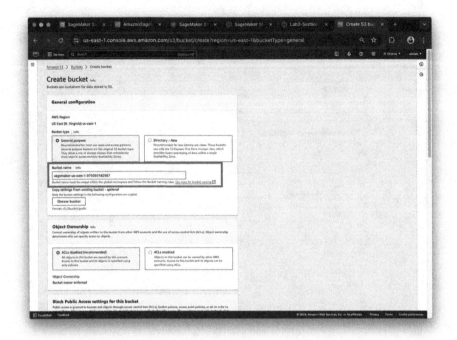

CHAPTER 6   BUILDING GENAI WITH SAP, LAMBDA, AND AMAZON BEDROCK

You can get your own AWS account number in the profile menu, as depicted in the following image:

CHAPTER 6 BUILDING GENAI WITH SAP, LAMBDA, AND AMAZON BEDROCK

Now you should see your bucket:

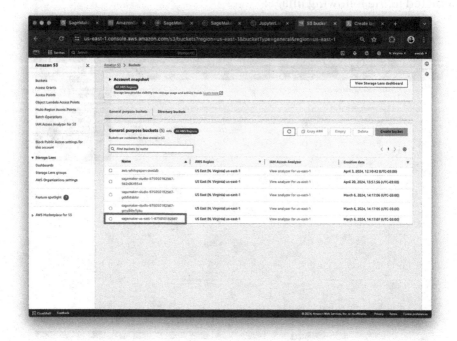

CHAPTER 6   BUILDING GENAI WITH SAP, LAMBDA, AND AMAZON BEDROCK

Step 10: Click the bucket.
Step 11: Click **Create folder**.

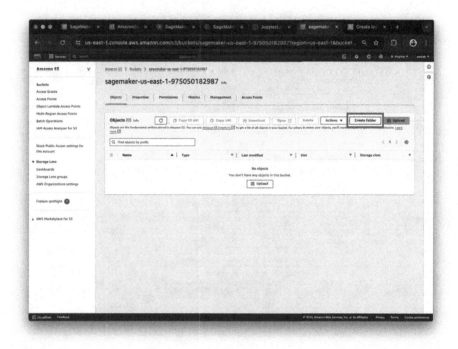

331

CHAPTER 6   BUILDING GENAI WITH SAP, LAMBDA, AND AMAZON BEDROCK

Step 12: Type "lambda_layer" for the folder name; then click **Create folder**.

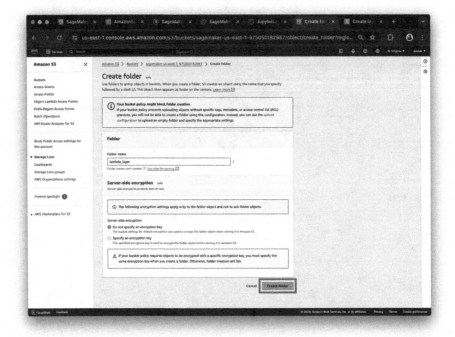

Now you can see that the folder exists:

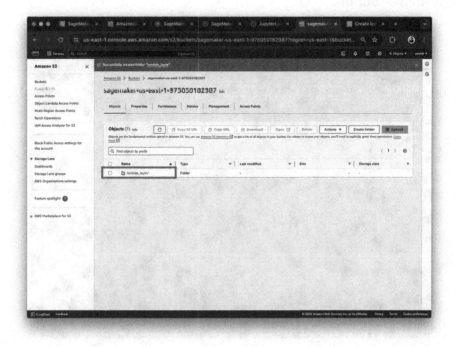

CHAPTER 6   BUILDING GENAI WITH SAP, LAMBDA, AND AMAZON BEDROCK

Step 13: Go back to the SageMaker Notebook and click the
$ (terminal) button.

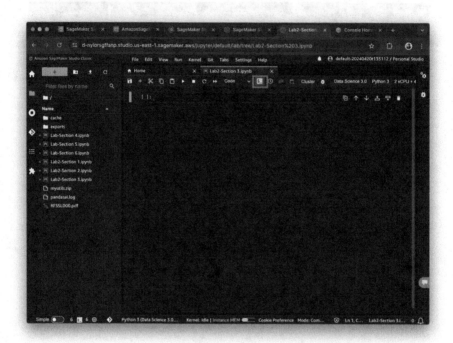

Step 14: Copy and paste the following code into a terminal and press Enter (remembering to change s3://sagemaker-us-west-2-AWSAccountNumber to the bucket you created in Step 9):

```
conda install -c conda-forge zip
/opt/conda/bin/pip install --upgrade -t ./python hdbcli pandas boto3
/opt/conda/bin/zip -r python.zip python
/opt/conda/bin/aws s3 cp python.zip s3://sagemaker-us-west-2-AWSAccountNumber/lambda_layer/
```

334

CHAPTER 6   BUILDING GENAI WITH SAP, LAMBDA, AND AMAZON BEDROCK

Here's an example:

```
conda install -c conda-forge zip
/opt/conda/bin/pip install --upgrade -t ./python hdbcli
pandas boto3
/opt/conda/bin/zip -r python.zip python
/opt/conda/bin/aws s3 cp python.zip s3://sagemaker-us-
east-1-975050182987/lambda_layer/
```

You should see a result like this if everything ran correctly and the file was copied to the S3 bucket:

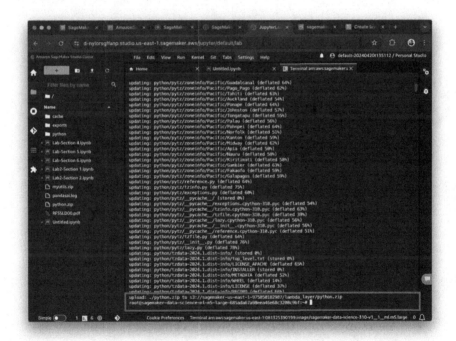

Step 15: Navigate to **AWS Lambda ➤ Layers**; then click **Create layer**. https://console.aws.amazon.com/lambda/#/layers

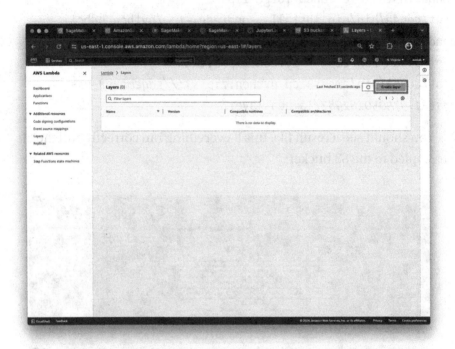

Step 16: Provide these values in the **Layer Configuration**:

- **Name**: llm
- **Description**: llm text to sql
- **Choose**: Upload a file from Amazon S3
- **Amazon S3 link URL**: s3://your-s3-bucket/lambda_layer/python.zip
- **Compatible architectures**: x86_64
- **Compatible runtimes**: Python 3.10

CHAPTER 6   BUILDING GENAI WITH SAP, LAMBDA, AND AMAZON BEDROCK

Remember to change *s3://sagemaker-us-west-2-AWSAccountNumber* to your own bucket created in Step 9.

Here's an example: *s3://sagemaker-us-east-1-975050182987/* `lambda_layer/python.zip`.

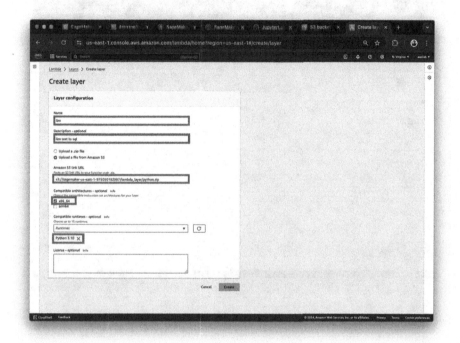

337

CHAPTER 6   BUILDING GENAI WITH SAP, LAMBDA, AND AMAZON BEDROCK

Step 17: Click the **Create** button.

If everything worked correctly, you should see this window:

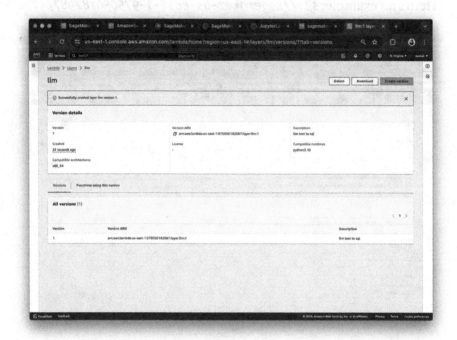

CHAPTER 6   BUILDING GENAI WITH SAP, LAMBDA, AND AMAZON BEDROCK

Step 18: Go to **AWS Lambda ➤ Functions**; then click **Create function.**
https://us-east-1.console.aws.amazon.com/lambda/home?region=us-east-1#/functions

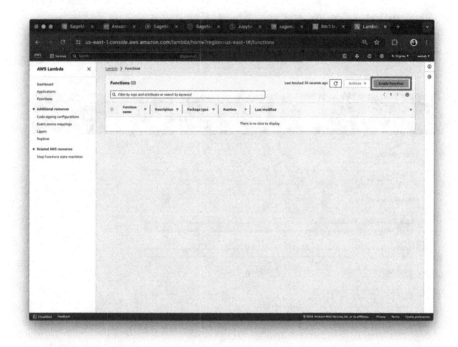

CHAPTER 6   BUILDING GENAI WITH SAP, LAMBDA, AND AMAZON BEDROCK

Step 19: Select **Author from scratch** and provide these values in the **Basic Information** area:

- **Function Name**: TextToSQL
- **Runtime**: Python 3.10
- **Architectures**: x86_64

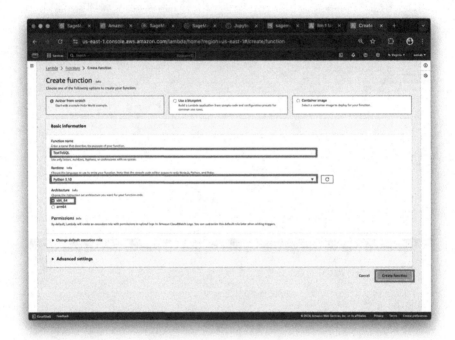

Step 20: Click the **Create function** button.

If everything is OK, you should see this window:

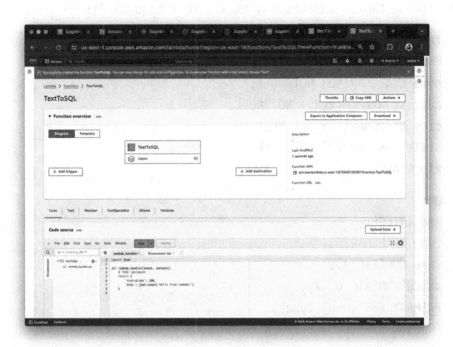

Step 21: Copy and paste the following code to `lambda_function.py`; then click the **Deploy** button (remembering to change the user, password, and SQL endpoint):

```
import json
import os
import sys
import boto3
import botocore
from hdbcli import dbapi
import pandas as pd
```

CHAPTER 6 BUILDING GENAI WITH SAP, LAMBDA, AND AMAZON BEDROCK

System = """Transform the following requests into valid SQL queries in SAP HANA dialect. Assume a database with the following tables and columns exists in schema USER1:

HOTEL:
  hno INTEGER PRIMARY KEY,
  name NVARCHAR(50) NOT NULL,
  address NVARCHAR(40) NOT NULL,
  city NVARCHAR(30) NOT NULL,
  state NVARCHAR(2) NOT NULL,
  zip NVARCHAR(6)

ROOM:
  hno INTEGER,
  type NVARCHAR(6),
  free NUMERIC(3),
  price NUMERIC(6, 2),
  PRIMARY KEY (hno, type),
  FOREIGN KEY (hno) REFERENCES HOTEL

CUSTOMER
  cno INTEGER PRIMARY KEY,
  title NVARCHAR(7),
  firstname NVARCHAR(20),
  name NVARCHAR(40) NOT NULL,
  address NVARCHAR(40) NOT NULL,
  zip NVARCHAR(6)

RESERVATION
  resno INTEGER NOT NULL GENERATED BY DEFAULT AS IDENTITY,
  rno INTEGER NOT NULL,
  cno INTEGER,
  hno INTEGER,
  type NVARCHAR(6),

```
  arrival DATE NOT NULL,
  departure DATE NOT NULL,
  PRIMARY KEY (
    "RESNO", "ARRIVAL"
  ),
  FOREIGN KEY(hno) REFERENCES HOTEL,
  FOREIGN KEY(cno) REFERENCES CUSTOMER

MAINTENANCE
  mno INTEGER PRIMARY KEY,
  hno INTEGER,
  description NVARCHAR(100),
  date_performed DATE,
  performed_by NVARCHAR(40)
```

Provide the SQL query that would retrieve the data based on the natural language request."""

```
#Setup boto3 client to access bedrock in a shared AWS Account
module_path = "../myutils"
#sys.path.append(os.path.abspath(module_path))
from myutils import bedrock, print_ww

# ---- ⚠ Un-comment and edit the below lines as needed for your AWS setup ⚠ ----
os.environ["AWS_DEFAULT_REGION"] = "us-east-1"
# os.environ["AWS_PROFILE"] = "<YOUR_PROFILE>"
# os.environ["BEDROCK_ASSUME_ROLE"] = "arn:aws:iam::<SharedAWSAccount>:role/Crossaccountbedrock"   # E.g. "arn:aws:..."

boto3_bedrock = bedrock.get_bedrock_client(
    assumed_role=os.environ.get("BEDROCK_ASSUME_ROLE", None),
    region=os.environ.get("AWS_DEFAULT_REGION", None),
    runtime=True
)
```

```python
def lambda_handler(event, context):
    event_body = json.loads(event["body"])
    print(event_body)
    human_input = event_body["prompt"]

    res = ask_question(human_input)

    return {
        "statusCode": 200,
        "headers": {
            "Content-Type": "application/json",
            "Access-Control-Allow-Headers": "*",
            "Access-Control-Allow-Origin": "*",
            "Access-Control-Allow-Methods": "*",
        },
        "body": json.dumps(res),
    }

def execute_query(sql_query):
    schema = "USER1"
    #Initialize your connection
    conn = dbapi.connect(
        address='<changehanaendpoint>.hana.trial-us10.
        hanacloud.ondemand.com',
        port='443',
        user='<changeuser>>',
        password='<changepassword>',
        encrypt=True,
        sslValidateCertificate=True
    )
    #If no errors, print connected
    #print('connected\n')
```

CHAPTER 6 BUILDING GENAI WITH SAP, LAMBDA, AND AMAZON BEDROCK

```python
    sql_result=pd.read_sql(sql_query,conn)
    #cursor = conn.cursor()
    #cursor.execute(sql_query)
    #rows = cursor.fetchall()
    #sql_result = rows
    #cursor.close()
    conn.close()

    return sql_result

def generate_natural_answer(question, result):
    str_list = ['Question: ', question, 'result: ', result]
    qna = ' '.join(str_list)
    body = json.dumps({
      "max_tokens": 600,
      "temperature": 0,
      "system": "Provide a concise answer for the Question and
      result",
      "messages": [{"role": "user", "content": qna}],
      "anthropic_version": "bedrock-2023-05-31"
    })
    response = boto3_bedrock.invoke_model(body=body,
    modelId="anthropic.claude-3-sonnet-20240229-v1:0")
    response_body = json.loads(response.get("body").read())
    result = response_body.get("content")
    final = result[0]['text']
    return final

def generate_sql_query(input_question):
    body = json.dumps({
      "max_tokens": 600,
      "temperature": 0,
      "system": System,
```

```python
        "messages": [{"role": "user", "content": input_
        question}],
        "anthropic_version": "bedrock-2023-05-31"
    })
    response = boto3_bedrock.invoke_model(body=body,
    modelId="anthropic.claude-3-sonnet-20240229-v1:0")
    response_body = json.loads(response.get("body").read())
    result = response_body.get("content")
    parts = result[0]['text'].split('````')
    sql_statement = parts[1][3:]
    return sql_statement

def ask_question(input_text):
    sql_query = generate_sql_query(input_text)
    #print(sql_query)
    query_result = execute_query(sql_query)
    #print(query_result)
    if len(query_result) <= 1:
        return generate_natural_answer(input_text, str(query_
        result))
    else:
        return query_result
```

CHAPTER 6   BUILDING GENAI WITH SAP, LAMBDA, AND AMAZON BEDROCK

Example: "USER1, Password1, 73098405-c4b1-437c-89cf-3248286c20fd"

After you copy the code into the `lambda_function`, your window should look like this:

CHAPTER 6   BUILDING GENAI WITH SAP, LAMBDA, AND AMAZON BEDROCK

After the deployment, you should see a message at the top saying that you successfully updated the textToSQL function.

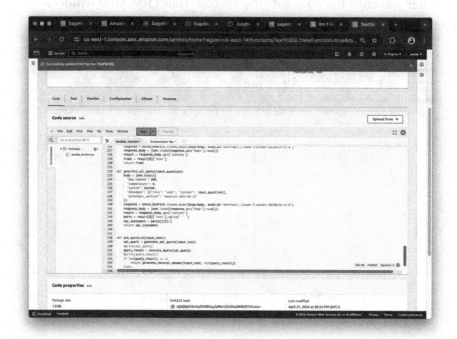

CHAPTER 6   BUILDING GENAI WITH SAP, LAMBDA, AND AMAZON BEDROCK

Step 22: Create the **myutils** folder.

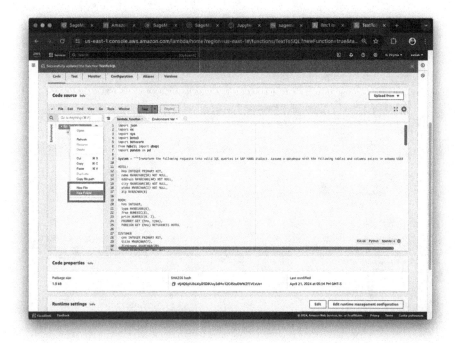

CHAPTER 6   BUILDING GENAI WITH SAP, LAMBDA, AND AMAZON BEDROCK

Step 23: Create the __init__.py file under the myutils folder.

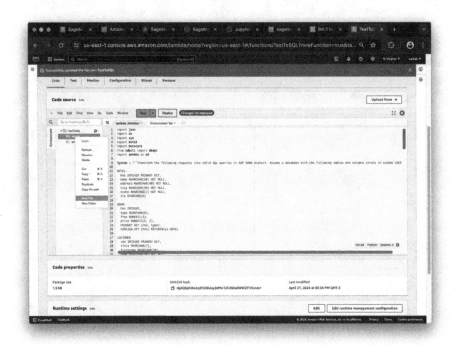

Step 24: Copy and paste the following code to file __init__.py; then click **Deploy**.

```
# Copyright Amazon.com, Inc. or its affiliates. All Rights Reserved.
# SPDX-License-Identifier: MIT-0
"""General helper utilities the workshop notebooks"""
# Python Built-Ins:
from io import StringIO
import sys
import textwrap
```

350

CHAPTER 6 BUILDING GENAI WITH SAP, LAMBDA, AND AMAZON BEDROCK

```
def print_ww(*args, width: int = 100, **kwargs):
    """Like print(), but wraps output to `width` characters
    (default 100)"""
    buffer = StringIO()
    try:
        _stdout - sys.stdout
        sys.stdout = buffer
        print(*args, **kwargs)
        output = buffer.getvalue()
    finally:
        sys.stdout = _stdout
    for line in output.splitlines():
        print("\n".join(textwrap.wrap(line, width=width)))
```

After the deployment, you should see a window like this:

CHAPTER 6   BUILDING GENAI WITH SAP, LAMBDA, AND AMAZON BEDROCK

Step 25: Create the bedrock.py file under the myutils folder.

Step 26: Copy and paste the following code to the file bedrock.py; then click **Deploy**.

```
# Copyright Amazon.com, Inc. or its affiliates. All Rights Reserved.
# SPDX-License-Identifier: MIT-0
"""Helper utilities for working with Amazon Bedrock from Python notebooks"""
# Python Built-Ins:
import os
from typing import Optional

# External Dependencies:
import boto3
from botocore.config import Config

def get_bedrock_client(
    assumed_role: Optional[str] = None,
    region: Optional[str] = None,
    runtime: Optional[bool] = True,
):
    """Create a boto3 client for Amazon Bedrock, with optional configuration overrides

    Parameters
    ----------
    assumed_role :
        Optional ARN of an AWS IAM role to assume for calling
        the Bedrock service. If not
        specified, the current active credentials will be used.
    region :
        Optional name of the AWS Region in which the service
        should be called (e.g. "us-west-2").
```

352

CHAPTER 6   BUILDING GENAI WITH SAP, LAMBDA, AND AMAZON BEDROCK

```
    If not specified, AWS_REGION or AWS_DEFAULT_REGION
    environment variable will be used.
runtime :
    Optional choice of getting different client to perform
    operations with the Amazon Bedrock service.
"""
if region is None:
    target_region = os.environ.get("AWS_REGION",
    os.environ.get("AWS_DEFAULT_REGION"))
else:
    target_region = region

print(f"Create new client\n  Using region: {target_region}")
session_kwargs = {"region_name": target_region}
client_kwargs = {**session_kwargs}

profile_name = os.environ.get("AWS_PROFILE")
if profile_name:
    print(f"  Using profile: {profile_name}")
    session_kwargs["profile_name"] = profile_name

retry_config = Config(
    region_name=target_region,
    retries={
        "max_attempts": 10,
        "mode": "standard",
    },
)
session = boto3.Session(**session_kwargs)

if assumed_role:
    print(f"  Using role: {assumed_role}", end='')
    sts = session.client("sts")
```

353

```python
        response = sts.assume_role(
            RoleArn=str(assumed_role),
            RoleSessionName="langchain-llm-1"
        )
        print(" ... successful!")
        client_kwargs["aws_access_key_id"] =
        response["Credentials"]["AccessKeyId"]
        client_kwargs["aws_secret_access_key"] =
        response["Credentials"]["SecretAccessKey"]
        client_kwargs["aws_session_token"] =
        response["Credentials"]["SessionToken"]

    if runtime:
        service_name='bedrock-runtime'
    else:
        service_name='bedrock'

    bedrock_client = session.client(
        service_name=service_name,
        config=retry_config,
        **client_kwargs
    )

    print("boto3 Bedrock client successfully created!")
    print(bedrock_client._endpoint)
    return bedrock_client
```

CHAPTER 6   BUILDING GENAI WITH SAP, LAMBDA, AND AMAZON BEDROCK

After the deployment, you should see a window like this:

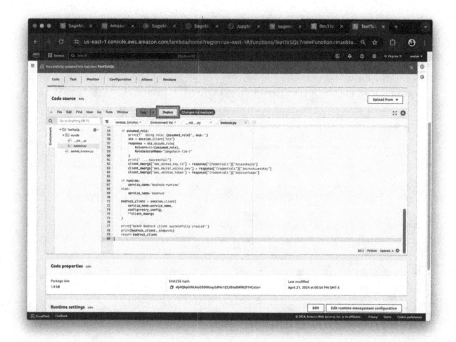

CHAPTER 6   BUILDING GENAI WITH SAP, LAMBDA, AND AMAZON BEDROCK

Step 27: Scroll the page up and click **Layers**.

CHAPTER 6 BUILDING GENAI WITH SAP, LAMBDA, AND AMAZON BEDROCK

Step 28: Click **Add a layer**.

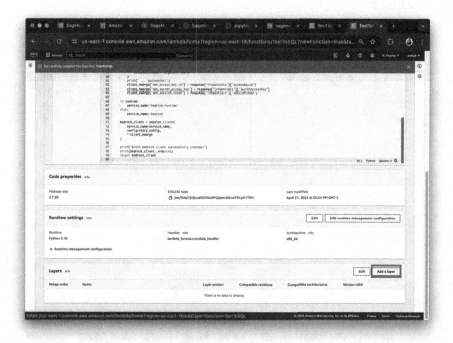

CHAPTER 6   BUILDING GENAI WITH SAP, LAMBDA, AND AMAZON BEDROCK

Step 29: Select available values in the **Choose a layer** area; then click the **Add** button.

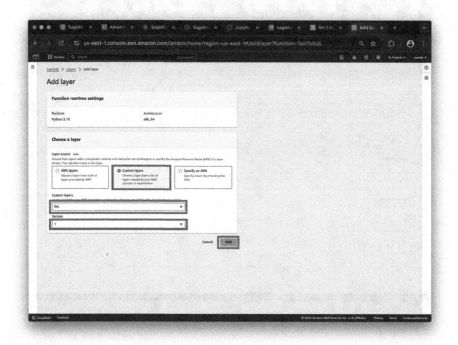

CHAPTER 6 BUILDING GENAI WITH SAP, LAMBDA, AND AMAZON BEDROCK

Step 30: Go to **Configuration** ➤ **General configuration** tab and click the **Edit** button.

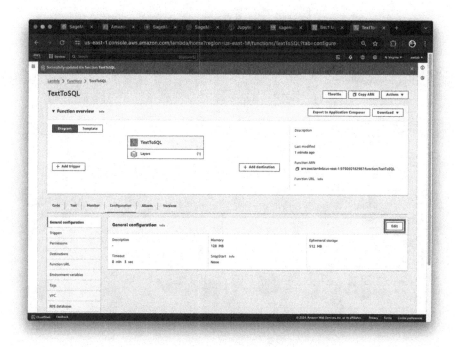

CHAPTER 6   BUILDING GENAI WITH SAP, LAMBDA, AND AMAZON BEDROCK

Step 31: Change the timeout from 3 seconds to **10 minutes**; then click **Save**.

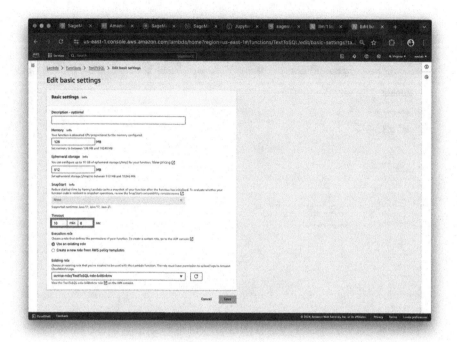

CHAPTER 6   BUILDING GENAI WITH SAP, LAMBDA, AND AMAZON BEDROCK

Step 32: Go to **Configuration** ➤ **Permissions**; then click **TextToSQL-role** as highlighted in the following image:

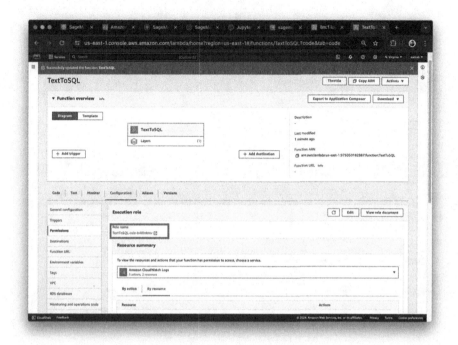

CHAPTER 6   BUILDING GENAI WITH SAP, LAMBDA, AND AMAZON BEDROCK

Step 33: Click the **Add permissions** drop-down and select **Create inline policy**.

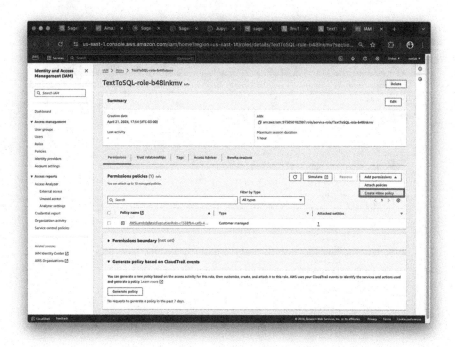

Step 34: Click **JSON**, copy and paste the following code to the **Policy editor**, and then click **Next:**

```
{
    "Version": "2012-10-17",
    "Statement": [
        {
            "Sid": "Statement2",
            "Effect": "Allow",
            "Action": [
```

362

CHAPTER 6   BUILDING GENAI WITH SAP, LAMBDA, AND AMAZON BEDROCK

```
                "bedrock:*"
            ],
            "Resource": [
                "*"
            ]
        }
    ]
}
```

Your window should look like this:

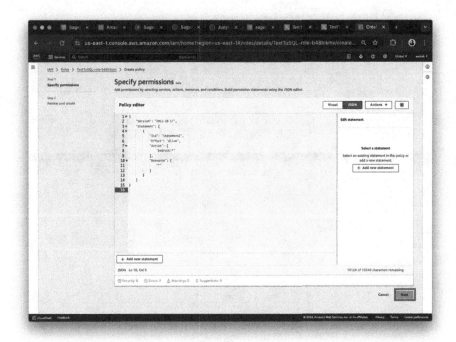

CHAPTER 6   BUILDING GENAI WITH SAP, LAMBDA, AND AMAZON BEDROCK

Step 35: Type "**bedrock-policy**" as the policy name; then click **Create policy**.

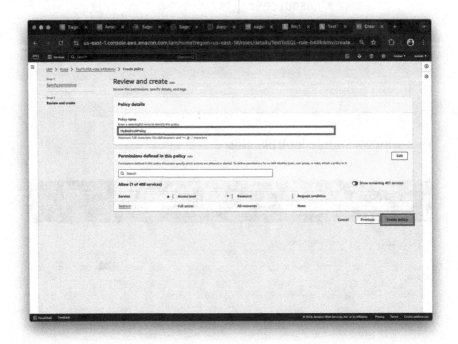

CHAPTER 6   BUILDING GENAI WITH SAP, LAMBDA, AND AMAZON BEDROCK

After you create the policy, you should see a window like this:

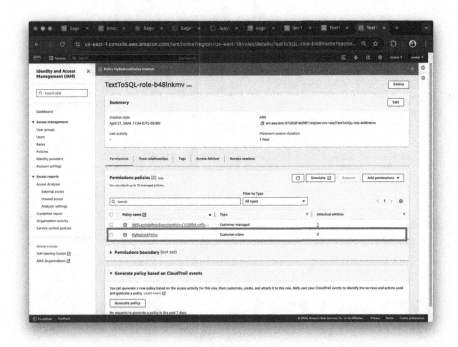

Step 36: Go to **AWS Lambda ➤ Testing**, type **"MyTestEvent"** in the Event Name field, and then copy and paste the following into **Event JSON**:

https://us-east-1.console.aws.amazon.com/lambda/home?region=us-east-1#/functions/TextToSQL?tab=testing

```
{
 "body": "{\"prompt\":\"How many Hotels are there in Seattle ?\"}"
}
```

CHAPTER 6 BUILDING GENAI WITH SAP, LAMBDA, AND AMAZON BEDROCK

Your window should look like the following example:

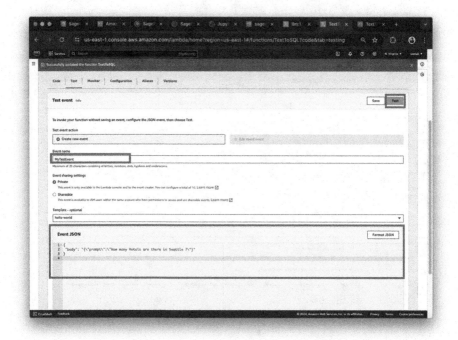

CHAPTER 6   BUILDING GENAI WITH SAP, LAMBDA, AND AMAZON BEDROCK

After a successful test, you will see a green message at the top of the window, as shown in the following example:

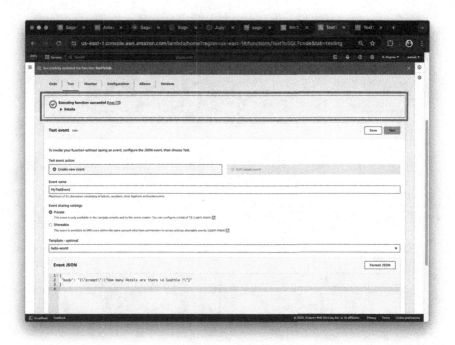

Remember, if you face errors, check your username, password, and SQL endpoint in `lambda_function.py`. Additionally, double-check if the SAP HANA Cloud instance is up and running.

CHAPTER 6   BUILDING GENAI WITH SAP, LAMBDA, AND AMAZON BEDROCK

Step 37 (Optional): Now, let's create an API so it can be executed from any system, such as SAP BTP Integration Suite. Click **Add trigger.**

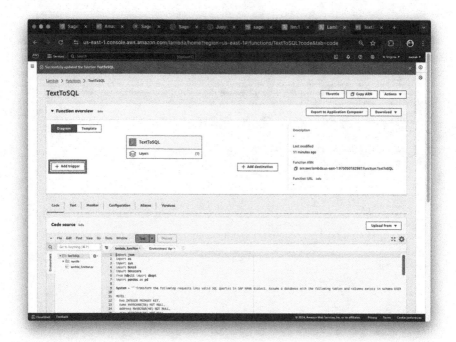

Step 38: Select **API Gateway**, click **Create a new API**, and then choose **REST API**.

Set the security mechanism to **API Key**; then click **Add**.

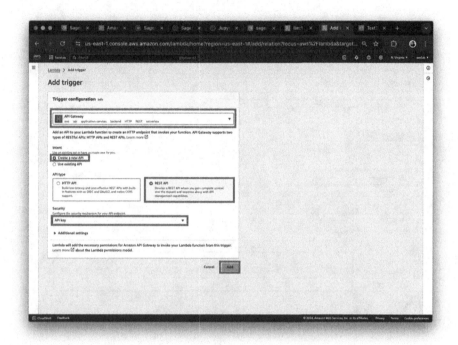

CHAPTER 6   BUILDING GENAI WITH SAP, LAMBDA, AND AMAZON BEDROCK

You will see a green message at the top of the window, as shown in the following sample, after the trigger creation. Observe that you now have an **API endpoint** and an **API key.**

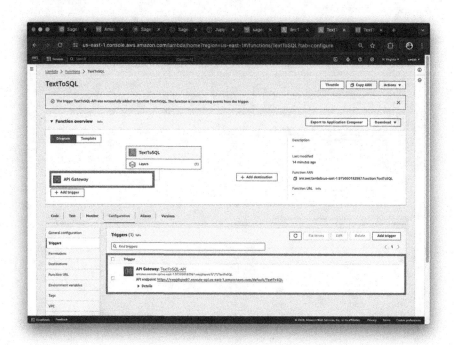

CHAPTER 6   BUILDING GENAI WITH SAP, LAMBDA, AND AMAZON BEDROCK

Step 39: Download **Postman**.
https://www.postman.com/downloads/
Step 40: Open Postman and click the **plus (+)** sign to open a new tab.

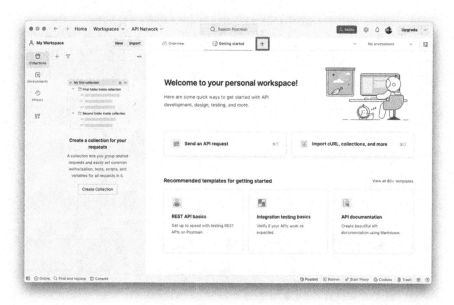

Step 41: Change the request to **POST** and provide the API endpoint, API key, and code.

- **URL**:

  https://xwpjdvpwb7.execute-api.us-east-1.amazonaws.com/default/TextToSQL

- **x-api-key**: wNDGg5iK6r4wIbcmADEPIaxw2ocuH5f65tOTDBFR

- **Body** (raw):

  { "prompt": "How many Hotels are there in Seattle?" }

371

CHAPTER 6 BUILDING GENAI WITH SAP, LAMBDA, AND AMAZON BEDROCK

Your window should look like the following:

CHAPTER 6   BUILDING GENAI WITH SAP, LAMBDA, AND AMAZON BEDROCK

Step 42: Click the **Send** button.
Your window should look like the following:

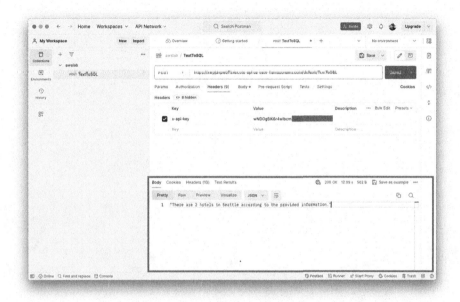

In this chapter, we explored some generative AI use cases for GenAI LLMs by implementing them with Amazon SageMaker Notebooks and SAP HANA Cloud. We leveraged Amazon Bedrock to dissect SAP reports with the help of summarization capabilities and employed Claude LLM as a financial analyst. We also discussed the features offered by PandasAI, SAP HANA Cloud, AWS Lambda, and API Gateway.

We are reaching the end of this journey, but there are some exciting and upcoming scenarios to discuss in the final chapter. Let's move on!

CHAPTER 7

The AI Journey Gets Started...

Generative AI has the potential to create significant business value along four major dimensions, including for SAP customers.

New Experiences

GenAI will assist in creating a conversational interface to access information and applications by employees and customers.

Productivity

GenAI will boost employee productivity by helping them create content, code, and images faster, with the use of tools such as Amazon Q Developer for coding.

CHAPTER 7   THE AI JOURNEY GETS STARTED...

Insights

The use of generative AI will result in an increased ability to extract insights from large volumes of documents, making it easier to share knowledge across an organization and make better, more informed decisions.

Creativity

Finally, generative AI will enhance creativity, often for product ideation, as well as product development.

Generative AI has the potential to transform how business is conducted, particularly in the context of SAP systems. SAP users may use AWS generative AI capabilities to automate SAP processes, discover insights, and expedite development. AWS generative AI capabilities can help make better decisions allowing users to analyze SAP data with Amazon Q in QuickSight, increase field productivity by summarizing complex documents with Amazon Bedrock and Amazon Q for Business, or solve business challenges with SAP products (such as SAP BTP, RISE with SAP, and SAP Datasphere) and AWS Services.

Generative AI has the ability to bring significant changes to the global economy. According to Goldman Sachs, generative AI might boost global GDP by 7% (or over $7 trillion) while also increasing productivity.

One of the major benefits of generative AI is the ability to interact with systems in new ways, often employing natural language similar to human conversation.

Changing how we connect with these systems will frequently lead to new opportunities, such as allowing users to get accurate information in more natural ways, engage in problem-solving discussions, and receive context-sensitive instructions.

Now that you're familiar with generative AI and Amazon Bedrock, I welcome you to look at business use cases that you could find useful in addressing your business challenges in a variety of scenarios.

CHAPTER 7   THE AI JOURNEY GETS STARTED...

Let's look at some promising uses for these capabilities to reimagine SAP-related operations.

There are various activities and roles across a company that may benefit from new AWS GenAI experiences—and the list is expanding as our teams apply these new technologies to more real-world scenarios. Examples include the following:

- *Finance managers* can now get financial insights in real time by querying about financial data and receiving replies from a conversational interface that uses Amazon Bedrock.
- *Sales inventory analysts* may use Amazon Bedrock and SAP Datasphere data to gain insights into key inventory indicators using natural language cues.
- *Legal or policy managers* may use document retrieval and Amazon Bedrock to assist in making sense of large amounts of documents to address legal and policy problems and better serve consumers and employees.
- *Shop floor employees* may utilize Amazon Q for Business to get critical details on manufacturing and maintenance operations from document repositories, including SAP and non-SAP, nonstructured data, using natural language.

Beyond these examples, generative AI may be used in various areas of business, including engineering, marketing, customer support, finance, and sales.

To gain a better understanding of possible use cases, let's consider the key challenges and potential benefits of employing generative AI in different professions.

CHAPTER 7   THE AI JOURNEY GETS STARTED...

Finance Manager

Managers across all enterprises want insights to prepare for meetings, keep informed when traveling, and quickly generate content based on these critical insights. This is especially true for financial managers, who are frequently tasked with providing such information. While static reports might be useful, they are often unable to keep up with the changing demands of many enterprises. Sometimes an answer is all that is needed.

Using Amazon Bedrock's AWS generative AI capabilities, financial managers can ask questions of their SAP and non-SAP data sources via a conversational interface, revealing new insights and providing the answers they need to drive their businesses forward. Generative AI can even summarize this information in easy-to-understand emails for stakeholders across the organization, helping build a common understanding around complex situations.

Similar approaches can be applied using different data within other functional areas and various industries such as healthcare, auto manufacturing, banking and financial services, telecommunication, and media and entertainment, where real-time insights and correspondence can be generated. No matter the functional role in question, it is worth considering how chatbots like these, leveraging an organization's own unique enterprise data, can help deliver insights and transform a business.

The use of generative AI in this field will have several main challenges:

- Long lead time to analyze data and uncover insights
- Inability to quickly relay resulting insights
- Ever-changing reporting requirements that add complexity to report development

CHAPTER 7   THE AI JOURNEY GETS STARTED...

Generative AI is likely to have the following benefits:

- Generates insights dynamically based on natural language prompting

- Provides near real-time insights from SAP data for a CFO or a finance team to accelerate decision-making

- Creates and shares summarized insights via email

Some examples of finance insights that a chatbot and NLP interface can provide to managers might be as follows:

- "Tell me the top five revenue regions globally."

- "Now tell me the top five revenue-generating territories within that top-performing region."

- "Lastly, please generate a summary that can be used in an email, summarizing the revenue attainment and comments on recent growth trends."

Beyond providing encounters and opportunities for growth, generative AI plays a role in enhancing efficiency on a comprehensive scale as well. We shall delve into how the utilization of AI features offered by AWS could elevate productivity for customers using SAPs services.

AI-generated systems can analyze data effectively and produce content autonomously in forms, like coding and task execution, showcasing its noteworthy potential to boost business growth significantly by aiding individuals as needed or through automated operations and workflows. Utilizing AI technology in processes such as SAP can lead to substantial productivity enhancements.

Let's look at a few of these examples:

- *Accounts payable managers* and *auditors* can increase their productivity by using generative AI tools like Amazon Bedrock to gain audit insights and detect anomalies in accounts payable transactions.

- *SAP ABAP developers* can leverage Amazon Bedrock to generate ABAP code snippets as well as ABAP documentation from a convenient SAP ABAP Assistance plugin for Eclipse IDE, helping speed up development efforts and future-proof code through proper documentation.

- *SAP system administrators* can significantly accelerate code generation with Amazon Q Developer to further automate system administration tasks.

- Lastly, *SAP enterprise leaders* can accelerate development and innovation with SAP BTP and AWS services using Amazon Q Developer. For example, code can be generated to render multilingual print forms with Adobe Print Service on SAP BTP and Amazon Translate.

Let's take a deeper look at two of these examples, starting with how Amazon Bedrock can help gain audit insights and detect accounts payable anomalies.

Accounts Payable Manager

Accounts payable managers face several challenges. Specifically, these managers must maintain compliance amid evolving requirements. To do so, they often must engage in time-consuming, tedious, and expensive

CHAPTER 7   THE AI JOURNEY GETS STARTED...

manual audit processes. In the process, they often need to access data from both structured data (like those in core databases) and nonstructured data (from individual invoice documents and other sources).

To address these challenges effectively as SAP users, consider streamlining the invoice processing workflow by utilizing AWS tools such as Amazon Textract and Amazon Translate for data extraction and translation purposes while leveraging Amazon S33 for storing data and gaining payment related insights through AWS Bedrock.

Automating invoice processing helps businesses accelerate the time it takes to approve payment, from receipt to payment approval, and decreases errors in the process leading to cost savings and improved customer experiences. Moreover, when combining AI technology with Amazon Bedrock, customers can extract insights from invoices and other documents with different formats and delegate complex tasks, through customized agents efficiently.

The following are the main challenges of generative AI in this area:

- Maintaining compliance with time-consuming audit processes
- Gathering insights from documents and nonstructured data

The following are the benefits of generative AI in this area:

- Creating audit assistants that leverage financial data
- Gaining insights on payments and receipts from invoices and other documents
- Creating agents to perform complex tasks on unique processes at scale

To support account payable managers, a kind of accounts payable invoice audit assistant can be provided using Amazon Bedrock with a dataset of invoice information.

Users can request an LLM model to identify trends or patterns in the services utilized and suppliers involved as the payment methods used in transactions. In response to this query, the model will present a breakdown of suppliers services procured and details on payment methods and their statuses. This information provides insights and context.

Furthermore, customers might request invoices that were overdue for payment; the system not only can present details of the invoices but also include information about the suppliers and possible reasons contributing to the delay.

This is just one simple example of on-demand assistance that can be provided by generative AI. These same types of inquiries can be deployed in automated workflows, constantly searching for similar review-worthy incidents and surfacing this information for accounts payable managers to act on ahead of dedicated audits.

SAP Developer

SAP customers have been embarking on digital transformation programs, many migrating to RISE with SAP. Many of those customers are currently running legacy SAP systems with complex, custom code that was built potentially over many years to address unique business process challenges or opportunities. While planning for their digital transformation projects, these customers need to continue maintenance and development in their legacy SAP systems, including this custom code. Also, as part of their transformation projects, customers are seeking to maintain a clean core based upon best practices built into SAP's S/4 HANA and RISE with SAP.

To accelerate and derisk transformation programs, customers need to understand custom ABAP programs in their legacy systems to analyze the impact on business processes and data. Understanding existing custom code and generating documentation can be challenging for several reasons:

- Lack of documentation for custom ABAP programs
- Unavailability of resources that built the custom code
- Lack of resources needed to review and understand the legacy custom code

This challenge can introduce risk, reduce productivity, and extend deadlines for projects.

With the AWS SAP ABAP Assistant using Amazon Bedrock, SAP customers can accelerate the delivery of new code for legacy systems by generating code snippets with natural language prompts. Both legacy support teams, as well as those involved in digital transformation programs for S/4HANA and RISE with SAP, are able to generate documentation on legacy and standard SAP code to accelerate transformation, reducing risk and increasing productivity. Resources can review this documentation to better understand business impact, helping to accelerate timelines and meet desired business outcomes.

The following are the main challenges of using GenAI as an ABAP developer:

- Productivity loss incurred searching for ABAP code
- Long lead time to analyze ABAP programs
- Lack of documentation of ABAP poses a risk to the business
- Advantages of utilizing GenAI as an ABAP developer
- Accelerates ABAP code development and documentation
- Improves the productivity of the ABAP developer
- Enables a path to a clean core and transformation

Chapter 7 THE AI JOURNEY GETS STARTED...

SAP customers can accelerate the delivery of new code for legacy systems by generating code snippets with natural language prompts using the AWS SAP ABAP Assistant (see Figure 7-1), a convenient plug-in for the Eclipse IDE that allows the functionality of Amazon Bedrock.

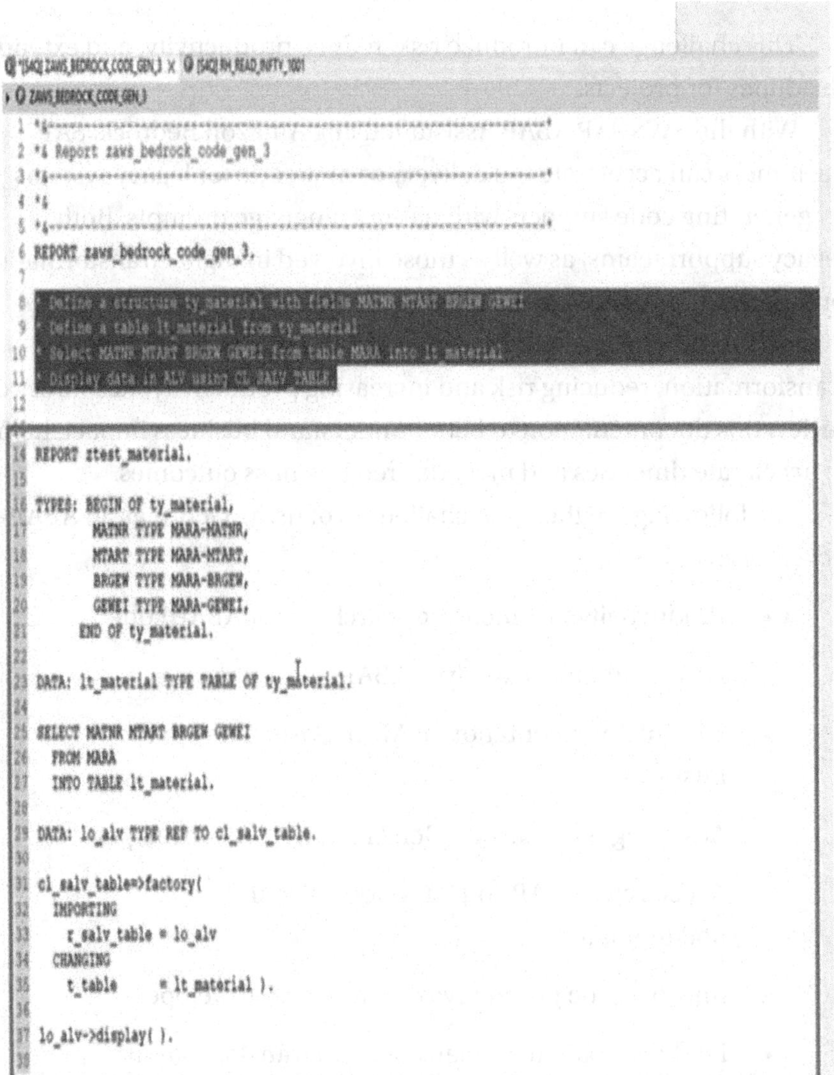

```
*&---------------------------------------------------------------------*
*& Report zaws_bedrock_code_gen_3
*&---------------------------------------------------------------------*
*&
*&---------------------------------------------------------------------*
REPORT zaws_bedrock_code_gen_3.

* Define a structure ty_material with fields MATNR MTART BRGEW GEWEI
* Define a table lt_material from ty_material
* Select MATNR MTART BRGEW GEWEI from table MARA into lt_material
* Display data in ALV using CL_SALV_TABLE


REPORT ztest_material.

TYPES: BEGIN OF ty_material,
         MATNR TYPE MARA-MATNR,
         MTART TYPE MARA-MTART,
         BRGEW TYPE MARA-BRGEW,
         GEWEI TYPE MARA-GEWEI,
       END OF ty_material.

DATA: lt_material TYPE TABLE OF ty_material.

SELECT MATNR MTART BRGEW GEWEI
  FROM MARA
  INTO TABLE lt_material.

DATA: lo_alv TYPE REF TO cl_salv_table.

cl_salv_table=>factory(
  IMPORTING
    r_salv_table = lo_alv
  CHANGING
    t_table      = lt_material ).

lo_alv->display( ).
```

Figure 7-1. SAP ABAP–generated code with Amazon Bedrock

CHAPTER 7   THE AI JOURNEY GETS STARTED...

Here we see how both legacy support teams, as well as those involved in digital transformation initiatives for SAP S/4HANA and RISE with SAP, are able to generate documentation on legacy and standard SAP code to accelerate transformation, reducing risk and increasing productivity (see Figure 7-2).

For anyone seeking to accelerate their development or documentation efforts, this functionality is a game changer.

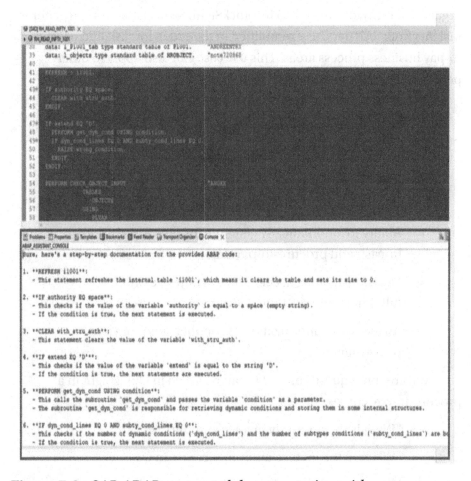

Figure 7-2. SAP ABAP–generated documentation with Amazon Bedrock

385

CHAPTER 7   THE AI JOURNEY GETS STARTED...

Now, let's dive deep into how SAP customers can extract insights from their unique enterprise information, making it easier to share knowledge across the organization and make better-informed decisions.

Order-to-Cash and Procure-to-Pay Insights

As we have seen, the need for instant insights is in high demand in current organizations. With Amazon Q in QuickSight, SAP customers can generate insights from SAP data, as exemplified in the order-to-cash and procure-to-pay business process areas. This use case can also apply to other personas depending on the SAP data available.

The following are the main challenges of GenAI in this area:

- Discrepancies in order fulfillment
- Inability to measure supplier spending and performance

The following are the benefits of generative AI in this area:

- Ability to automatically generate insights from order-to-cash and procure-to-pay data
- Decision-making acceleration to improve order fulfillment and on-time delivery rate
- Analysis and optimization of supplier spending and performance

Here are some questions that Q can be asked in QuickSight in a procure-to-pay context:

- "What is the quarterly invoice amount?"
- "Why did the invoice amount increase in Q1?"
- "What were the top five suppliers by Net price during this time?"
- "What is the breakdown by spend category?"

CHAPTER 7   THE AI JOURNEY GETS STARTED...

Here are some questions that Q can be asked in Q in QuickSight in an order-to-cash context:

- "Who are the top three customers with the highest number of ordered items?"
- "What are the total sales by Material group?"

As noted previously, this incredible "generative BI" capability can be used to help SAP users get insights and make better-informed decisions faster, including for key processes like order-to-cash and procure-to-pay processes.

Figure 7-3 shows an example of the dashboard and visualization of SAP data that can be generated with Amazon Q in QuickSight.

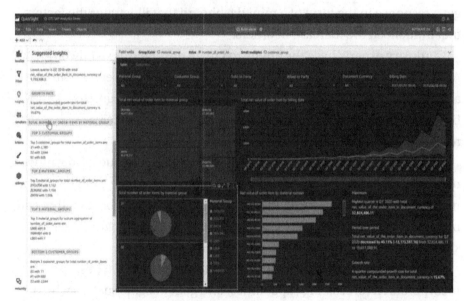

Figure 7-3. Amazon Q QuickSight dashboard example

Order-to-cash and procure-to-pay analysts can generate dashboards, accelerate analysis by easily creating calculations, and even develop sharable data stories, all with natural language.

387

CHAPTER 7   THE AI JOURNEY GETS STARTED...

AI-Powered Dashboard Authoring Experience

Amazon QuickSight is the ML-powered business intelligence service that has Amazon Q features built in.

Amazon QuickSight allows the use of natural language to quickly build visuals for dashboards and reports, build calculations without looking up or learning specific syntax, and quickly update visuals by describing any desired changes.

With Amazon Q in QuickSight, users can ask their dashboards questions like "Why did the number of orders increase last month?" and get a response in the form of a mini dashboard that illustrates the factors that influenced the increase in orders.

Industry experts have the ability to create dashboards swiftly on Amazon Q by articulating the data they want to see represented. Simply ask a query on Amazon Q such as "Display sales data categorized by region, per month using a stacked bar chart." You'll receive a representation promptly within seconds.

Amazon Q users on QuickSight have the option to request a narrative detailing the business changes over the month for a review with leaders in no time at all. Amazon Q will then generate a story based on data that can be securely shared within the organization to facilitate alignment among stakeholders and enhance decision-making processes.

Amazon Q in QuickSight allows users to change chart types, incorporate AI forecasting capabilities, and gain insights such as which material groups are the top performers.

It might already be clear that generative AI productivity capabilities support multiple lines of business, but generative AI's breakthrough creative capabilities can also be applied in areas like product ideation, product development, and other key functions.

AWS generative AI capabilities can be used for product description and image ideation for your SAP systems, including your master data, marketing, and product development teams.

- *Master data managers* can use Bedrock and Fiori apps to help provide product descriptions in multiple languages.
- Marketing managers can generate images for sales and marketing materials with SAP and non-SAP data.
- Product development managers can generate images and product alternatives for prototyping with SAP and non-SAP data.

SAP Business AI

The SAP Business AI goes beyond hype to assist customers in showcasing their potential by integrating advanced AI advancements directly into the frameworks that drive a significant portion of the global business landscape artificial intelligence is seamlessly woven throughout SAP's complete range of offerings.

SAP has announced extensive new integrations of natural language generative AI copilot Joule, unveiling a swathe of new Business AI capabilities throughout their comprehensive portfolio of business applications.

By quickly sorting and contextualizing data from multiple systems to surface smarter insights, Joule helps people get work done faster and drives better business outcomes in a secure, compliant way.

Joule was announced in September 2023 with initial integrations in SAP SuccessFactors and SAP Start. Since then, SAP has integrated Joule into SAP S/4HANA Cloud Public Edition and SAP S/4HANA Cloud

CHAPTER 7   THE AI JOURNEY GETS STARTED...

Private Edition, as well as in multiple products, including SAP Customer Data Platform, SAP BTP Cockpit, SAP Build, SAP Build Code, and SAP Integration Suite (see Figure 7-4).

Figure 7-4. *Joule in the supply chain running*

In the second half of 2024, Joule is also slated to be available in SAP Ariba, SAP Analytics Cloud, and multiple supply chain management solutions from SAP. Additionally, Joule will be able to understand and respond to queries in a growing number of languages, including German, Spanish, French, and Portuguese, making it even more useful for global organizations.

Further integration of Joule across the SAP enterprise portfolio will include the following arenas.

ERP and Finance

Customers can use Joule to optimize financial processes in various SAP solutions:

- Billing specialists can use Joule with SAP Advanced Financial Closing to unlock the unstructured content of inbound customer correspondence for collections, improving the accuracy of risk determinations. Joule can also generate a summary of project profitability, prioritizing projects for further analysis and proposing actions for project improvement.

- SAP Fiori applications now benefit from Joule-enabled insights, with the copilot able to summarize available data and produce visual data narratives.

Human Resources

People are able to utilize Joule to receive responses to questions regarding their company's HR policy papers in the SAP SuccessFactors system, which saves time and improves the staff satisfaction level. Moreover, Joule offers an interactive method for employees to inquire about topics and carry out HR tasks linked to their individual HR and talent information.

Joule's document grounded function offers responses using business documents from SAP and other third-party sources like Microsoft SharePoint. This feature enhances Joule's ability to provide answers based on customers' structured and unstructured business data.

Industries

Commodities traders and business operators will be spared from repetitive data-entry tasks in the SAP S/4HANA Cloud Private Edition, a solution for commodity management for physical contracts. With Joule, business operators can create new commodities deals by simply describing the deal using conversational language, giving them more time to focus on producing better commercial outcomes for their clients. This capability is planned to be available in the second half of 2025.

IT and Platforms

Business analysts, IT professionals, and process owners can rapidly automate business processes at scale using generative AI capabilities made possible by Joule. New capabilities that support enterprise automation include the following:

- Joule copilot integration with SAP Build Process Automation, which allows customers to generate automation and workflows using natural language requests in addition to the standard no-code visual editor.

- Joule copilot integration with SAP Signavio, which supports natural language requests to generate process models in addition to the manual process editor.

- Customers using SAP Integration Suite can create integration flows and connect to any application using simple natural language with Joule.

CHAPTER 7   THE AI JOURNEY GETS STARTED...

- Compliance and IT professionals will be able to leverage Joule with the regulatory change manager tool, a cloud application built on SAP BTP. This tool evaluates a vast number of regulatory updates, putting them into the context of a customer's business and their SAP solutions. It also provides impact analysis across SAP products and solutions to help customers maintain compliance and run their day-to-day business without disruption.

- SAP is introducing ABAP Developer capabilities in Joule that let software developers write ABAP code using generative AI. The first feature available for beta release and planned for general availability in the second half of this year lets software developers create ABAP business objects using generative AI in a SAP BTP/ABAP environment.

- Additional ABAP Developer capabilities in Joule, planned for early 2025, will be available for SAP S/4HANA Cloud Public Edition, SAP S/4HANA Cloud Private Edition, and SAP BTP/ABAP environments.

- Additionally, to help developers generate, complete, and test ABAP code efficiently, SAP is fine-tuning a model on its proprietary ABAP code, leveraging NVIDIA NeMo microservices.

- Joule-powered SAP Build Code now lets customers be more agile by quickly building Fiori front ends to their SAP S/4HANA Cloud systems using generative AI, simplifying SAP S/4HANA Cloud extensions.

Sourcing and Procurement

Sourcing managers can create requests for proposals from suppliers or vendors much faster in SAP Ariba Sourcing, thanks to intelligent product and supplier recommendations from Joule. These recommendations consider cost-effectiveness, carbon footprint impact, local compliance regulations, and past transactions. This capability is planned to be available in the second half of 2025.

Supply Chain

Supply chain planners can interact with SAP Integrated Business Planning for Supply Chain by relying on Joule's detailed analysis of supply-chain planning runs. This analysis will identify the root causes behind delays in filling orders and suggest corrective measures.

Logistics providers can use Joule to speed up the process of receiving goods and planning transportation in SAP S/4HANA Supply Chain for transportation management.

Generative AI Hub

By utilizing the reference architecture (JRA; see Figure 7-5), SAP clients can speed up the implementation of AI and update crucial business procedures based on SAP solutions. The advancements can be applied in integrated scenarios within RISE with SAP and the intelligent scenario lifecycle management feature as an integrating element or independently on SAP BTP. Customers have the option to utilize the Generative AI Hub in SAP AI Core and AWS services to develop AI solutions enhancing custom AI capabilities across SAPs range of cloud solutions and applications.

CHAPTER 7 THE AI JOURNEY GETS STARTED...

This collaboration aims to bring perspectives and improvements to areas of business operations, like finance and human resources, among others.

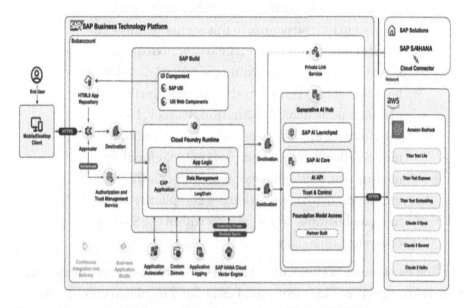

Figure 7-5. Joint Reference Architecture with SAP BTP and Amazon Bedrock

To help you grasp the concepts, here is an overview of this structure:

Amazon Bedrock: Customers have the option to access a variety of language models (LLMs) through APIs in this context known as JRA (Joint Research Agreement). We utilize Amazon Titan and Anthropic's Claude in our work. Both are a set of models pre-trained by AWS using data sets to serve as versatile and robust models suitable for various applications. Users can employ these models as they are or tailor them to their needs by incorporating their data privately.

Generative AI Hub in SAP AI Core: The SAP AI Core service offers customers access to AI resources like language models and provides a cohesive interface for SAP applications within the SAP BTP ecosystem infrastructure. In this scenario outlined in the JRA documentation, the Generative AI hub serves as a component within the SAP AI Core by handling access control and overseeing the lifecycle of interactions related to Amazon Bedrock. It acts as a point for applications to connect and utilize models effectively. SAP employs content filtering measures and risk management protocols through the Generative AI Hub to ensure compliance and mitigate business and legal risks within the SAP ecosystem on a scale.

The Generative AI Hub within SAP AI Core provides developers with access to a variety of LLMs from sources under a regulated commercial and legal structure framework. By using this platform developers can coordinate models efficiently and effectively. Moreover, the Generative AI Hub integrates with the vector capabilities in SAP HANA Cloud to assist developers in minimizing model hallucinations and integrating data as embeddings to offer tailored outcomes for specific scenarios.

Data storage and retrieval: SAP HANA Cloud serves as a versatile database management system for enabling the creation and implementation of data applications while the SAP HANA Vector engine integrates retrieval augmented generation (RAG) enhancing the performance of LLMs.

Application development: The Cloud Application Programming (CAP) model is a method for creating cloud applications utilizing SAP Build; it offers an efficient approach for data modeling with services seamlessly integrated into it. CAP gives developers access to source and SAP frameworks, facilitating faster development and fostering innovation opportunities efficiently in a project. CAP serves as the entity layer of the application, complemented by a SAP UI front end.

Authentication and identification: Effectively utilizing language models to their potential while ensuring compliance with user permissions during queries presents a significant integration hurdle to overcome. To address this challenge successfully and streamline the management of both SAP and SAP identity lifecycles for model prompting purposes, we propose leveraging SAP Business Technology Platform (BTP) in conjunction with SAP Identity Provisioning services.

With the consolidated use of these components, customers will now be able to build a scalable and reliable full-stack generative AI-powered SAP application leveraging Amazon Bedrock and SAP BTP services. This JRA pattern not only can be adapted to a wide range of business process extensions within an enterprise-grade SAP landscape, but it can also be helpful in maintaining a SAP-recommended clean core approach.

CHAPTER 7   THE AI JOURNEY GETS STARTED...

SAP Interactive Value Journey

SAP has enabled SAP Interactive Value Journey as a central repository for use cases to help customers unlock AI's true potential, which hinges on not only envisioning bold possibilities but also making them a reality to achieve real-world results. The aim is helping to bring out their best with business AI.

SAP Interactive Value Journey brings many use cases into a playlist format, covering AI value journeys in Cloud ERP, spend management and business network, customer experience, HR, BTP, and business AI (see Figure 7-6).

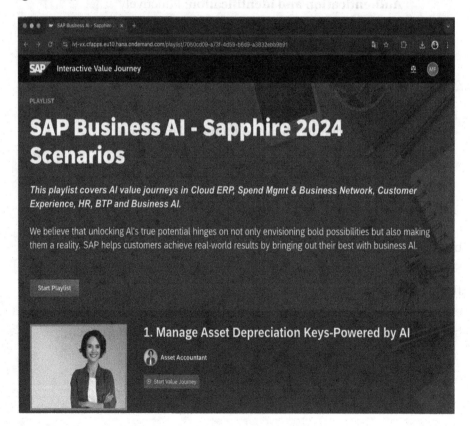

Figure 7-6. *SAP Interactive Value Journey*

CHAPTER 7   THE AI JOURNEY GETS STARTED…

Figure 7-7 refers to the use case of adopting AI technologies to boost productivity.

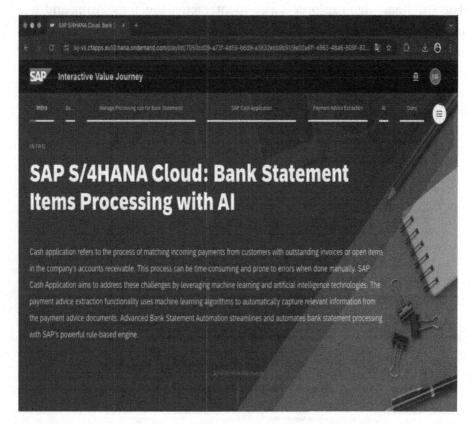

Figure 7-7. *SAP S/4HANA Cloud: bank statement items processing with AI*

The initiative is a good approach to familiarize individuals and company decision-makers with how to materialize AI benefits toward business challenges by supporting business needs.

CHAPTER 7   THE AI JOURNEY GETS STARTED...

SAP Road Map Explorer

The SAP Road Map Explorer is a go-to interactive tool for viewing and customizing road maps according to a customer's needs. To use it, just log on at the https://roadmaps.sap.com/ with a SAP Universal ID. Figure 7-8 shows the SAP roadmap for AI and upcoming features.

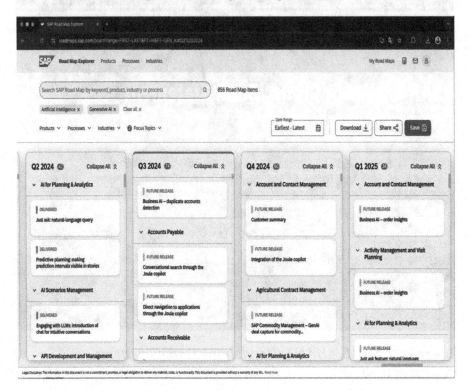

Figure 7-8. SAP Road Map Explorer

Going Beyond with Generative AI

Let me share some examples to demonstrate how combining AWS and SAP technologies can effectively address business challenges in real life scenarios instead of just theory.

CHAPTER 7 THE AI JOURNEY GETS STARTED...

Transforming Manufacturing with AWS and SAP

This case study is centered on the manufacturing industry applications of AI using SAP settings to illustrate the advantages of AI and connect the theoretical ideas I've discussed with real-world results.

Let's take a look at a real-life example involving a manufacturing firm that encountered obstacles related to delays in production and ineffective handling of stock levels. They tackled these issues by combining AWS tools like Amazon QuickSight with SAP Datasphere system to enhance the order-to-cash flow, which led to a 30% decrease in delays and better management of their supply chain. Access to updates on production and stock data empowered managers to make decisions based on data insights; this resulted in shorter lead times and boosted overall operational effectiveness. The collaboration showcases the potential for productivity enhancements by merging AWS's AI with SAP systems (see Figure 7-9).

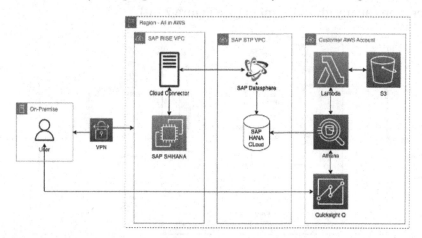

Figure 7-9. Insights using natural language in Amazon Q in QuickSight with SAP Datasphere

CHAPTER 7   THE AI JOURNEY GETS STARTED...

Working with SAP and AWS

It is easier to understand processes when discussing how advanced technologies such as SAP and AWS work together seamlessly. I believe that the visuals I included can clarify how information moves between systems and make the material easily understandable. I hope this clarifies things for individuals who might not be well versed in the specifics of cloud services and artificial intelligence tools.

To better understand how AWS services interact with SAP systems, Figure 7-10 outlines a typical workflow. Data from SAP S/4HANA is processed through AWS Bedrock for insights generation, leveraging Amazon Textract to extract relevant information from nonstructured data sources, and Amazon Q in QuickSight is used for real-time visualization. This workflow shows how users can query both structured and unstructured data seamlessly, providing a holistic view of business operations at any given moment (see Figure 7-10).

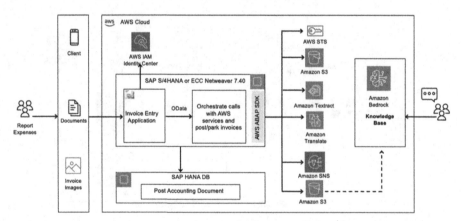

Figure 7-10. SAP Accounts payable process with AWS AI/ML

CHAPTER 7   THE AI JOURNEY GETS STARTED...

Exploring the Strategic Value of Generative AI Beyond Productivity

Productivity is a key benefit of generative AI, and I'd like to explore its strategic value to highlight its broader potential for business transformation. By emphasizing creativity, product development, and customer experience, this section will appeal to a wider audience, including business leaders looking for innovative ways to drive growth. Based on these observations, generative AI should be seen as a long-term investment rather than just a short-term productivity tool.

Generative AI does more than boost productivity—it also plays a pivotal role in driving innovation and strategic growth. In product development, for example, generative AI can assist in ideating new product concepts by generating multiple variations of designs based on market data or customer preferences. Marketing teams can leverage AWS generative AI tools to create personalized campaign content tailored to specific demographics, driving higher engagement rates. These creative applications of AI can enable businesses to differentiate themselves in a competitive marketplace.

Tangible Effects on Financial Operations

Generative AI and AWS offer advantages that are clearly demonstrated through the return on investment they provide in terms of cost savings and efficiency enhancements with direct business benefits from adopting these innovations. The credibility stemming from the profit benefits should appeal to decision-makers seeking rationale to embrace these emerging technologies.

In one implementation project, an international financial institution leveraged Amazon Bedrock to streamline their accounts procedures resulting in a 40 % decrease in manual tasks and a reduction of payment processing time from 10 days to 5 days. This initiative not only led to operational cost savings of $500,000 for the company but also enhanced vendor delight by guaranteeing prompt payments. These quantifiable results highlight the return on investment that enterprises can anticipate from integrating AWS machine learning solutions into their SAP setups.

Connecting Audiences Using a Prompt

Imagine having the ability to inquire about your SAP system's revenue-generating products from the quarter and getting a comprehensive real-time report filled with practical insights within seconds. Consider the amount of time this could save for your finance and sales teams alike. Innovation enterprises have already incorporated AWS AI capabilities into SAP systems to avail of this technology. How does your organization plan to make the most of these capabilities to stay ahead of the game?

Predicting with Generative AI

As artificial intelligence advances further in its development journey, its significance in analytics will continue to grow. Think about a scenario where a system not only looks at sales data but also forecasts upcoming trends to assist organizations in refining their production timelines and preparing for spikes in demand. Through the combination of AWS's AI technology and SAP systems, enterprises can transition from making decisions based on reactions to adopting approaches that set themselves up for sustained prosperity. Along with these progressions arise difficulties, though, such as the necessity for AI oversight and safeguards for data security. These will be significant areas of emphasis in the coming years.

CHAPTER 7   THE AI JOURNEY GETS STARTED...

Redefining Customer Satisfaction

Companies can greatly enhance their customer service departments by combining AWS AI with SAP systems. Through the utilization of natural language processing (NLP) technologies such as Amazon Lex, businesses are able to develop chatbots that deliver assistance to customers. These chatbots have the capability to retrieve real-time SAP customer information delivering tailored suggestions and resolutions based on the customer's interactions and preferences (see Figure 7-11). This results in a more gratifying customer journey ultimately aiding companies in bolstering customer satisfaction and loyalty.

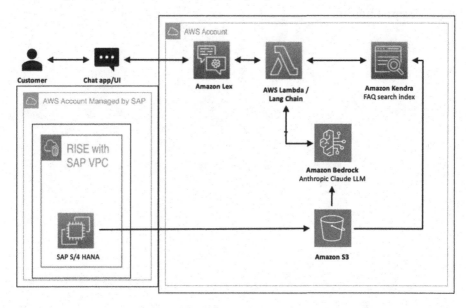

Figure 7-11. Enhancing the customer experience with AI

CHAPTER 7   THE AI JOURNEY GETS STARTED...

Last, but Not Least

Congratulations! You have completed this book, and I hope you have enjoyed the journey. As I write these last words, I have total confidence that AI will be continually changing, and large language models will be getting more and more specialized in certain tasks to support a variety of demands.

As organizations and users begin to embrace AI, it is expected that they both will achieve higher productivity, engagement, and growth.

Let's stay curious and pay attention to the AI evolution and the effects that it will bring us. Thank you for embarking on this journey with me, which has just begun.

Index

A

AGI, *see* Artificial general intelligence (AGI)
AI, *see* Artificial intelligence (AI)
AlphaGo system, 13
Amazon, 32
Amazon Bedrock, 33, 126
    advanced image playground, working, 159–161, 164, 165, 167–172
    application authorization page, 195, 196
    AWS account, login, 127, 131–133, 214, 216, 217
    chat playground, working, 133, 135, 138, 140–142, 144–146
    copy SQL endpoint, 207–209
    default identity provider, 193
    default values, 199, 200, 202
    HANA, 180
    image playgrounds, working, 146, 148, 150–157, 159
    instance name and password, 198
    IP addresses, 201
    Python language, 173
    review/create button, 203, 204
    SAP BTP Cockpit, 178, 206
    SAP BTP Trial account, create, 174, 177
    SAP HANA Cloud, 189, 191, 197
    SAP HANA Cloud instance, 182
    SAP HANA Database, 192
    security, 184
    SQL code block, 211, 213, 214
    SQL console session, 210
    subaccounts/directories list, 179
    tools plan, 183
    trial account name, 181, 190
    user email and password, 194
    user role assignments panel, 186–188
Amazon Lex, 405
Amazon Q, 81
Amazon Q Developer, 94
Amazon QuickSight, 388
Amazon SageMaker Notebook, 173
Amazon SageMaker Studio
    add permissions, 226, 227
    AWS console, open, 218
    code, 227, 228
    domain settings, 223
    execution role, 224–226

INDEX

Amazon SageMaker Studio (*cont.*)
    IAM console, 224
    JSON policy editor, 227
    launch, 236
    Open Studio, 231–236
    policy, 229
    setup single user, 220–222
Amazon SageMaker Studio Notebook, 248
Amazon Titan, 285
Amazon Titan G1-Express model, 140
Amazon Titan Stable Diffusion FM model, 159
Amazon Titan Text G1-Express model, 146
ANNs, *see* Artificial neural networks (ANNs)
ARIMA, *see* Autoregressive integrated moving average (ARIMA)
Artificial general intelligence (AGI), 55, 85
Artificial intelligence (AI)
    advanced technologies, 8–11
    AGI, 12
    Amazon AML, 32, 33
    Amazon Bedrock, 33–36, 38–41
    Amazon EC2 Inf2, 42
    attention, 16, 17
    AWS's custom Trainium chip, 42
    BERT, 22
    ChatGPT, 31
    encoders/decoders, 17, 18
    foundation models, 18
    Gemini, 23
    generative AI, 13
    Google BARD, 23
    Google BARD *vs.* Gemini, 24
    LLMs, 14, 15
    machines, 5
    Midjourney, 26, 27
    OpenAI, 27, 28
    OpenAI GPT series, 29, 30
    OpenAI's GPT, 25, 26
    R&D, 6
    SAP AI foundation, 48, 50–52
    SAP build code, 44, 46–48
    SAP joule, 42–44
    traditional ML *vs.* foundation models, 18, 19, 22
    transformers, 15, 16
    turing test, 5
    voice recognition, 1
Artificial neural networks (ANNs), 7
Autoregressive integrated moving average (ARIMA), 91
AWS Key Management Service (KMS), 37
AWS's AI technology, 404

B

Bidirectional long short-term memory (Bi-LSTM), 91
Bi-LSTM, *see* Bidirectional long short-term memory (Bi-LSTM)

INDEX

Business cyborg, 99
Business values, Gen AI
   accounts payable
      managers, 380–382
   AI-powered dashboard,
      388, 389
   creativity, 376, 377
   experience, 375
   finance manager, 378–380
   order-to-cash/procedure-to-
      pay insights, 386, 387
   productivity, 375
   SAP developer, 382, 383,
      385, 386

C

Chatbots, 57, 87
ChatGPT, 9, 31, 94, 95
Cloud Application Programming
   (CAP) model, 397
CNNs, *see* Convolutional neural
   networks (CNNs)
Computer vision, 4
Continuous integration/
   continuous deployment
   (CI/CD) process, 123
Convolutional neural networks
   (CNNs), 87
Copilot, 94
COVID-19 pandemic, 57, 58, 88
Cutting-edge deep learning
   method, 70
Cyborg, 98

D

DD, *see* Deep Docking (DD)
Deep Docking (DD), 92
Deep learning, 2, 87
Drug repurposing, 91

E

EC, *see* Evolutionary
   computation (EC)
ESs, *see* Expert systems (ESs)
Evolutionary computation (EC), 7
Expert systems (ESs), 7
Explainable AI (XAI)
   techniques, 124

F

FLAN T5, 285
FMs, *see* Foundation
   models (FMs)
Foundation models (FMs), 133

G

GANs, *see* Generative adversarial
   networks (GANs)
Gemini, 23, 81
GenAI, *see* Generative AI (GenAI)
Generative adversarial networks
   (GANs), 81
Generative AI (GenAI), 13, 55
   AWS/SAP, 401, 402
   ChatGPT

INDEX

Generative AI (GenAI) (*cont.*)
   Amazon Q, 81, 82
   conversational AI
      interfaces, 75
   Gemini, 81
   Microsoft, 78–80
   Microsoft Copilot, 80
   model evolution, 71, 72, 74
   pop culture, 76
   pop culture influence, 75
   transformer-based
      architecture, 70
consistency/trust, 118, 119
customer satisfaction, 405
developers, 114
ethical implications, 117, 118
financial operations, 403, 404
human developers, 119–121
industries, reshaping, 55, 56
LLMs, 112, 113
open-source licensed
      programs, 115
pandemic
   capital goods
      industry, 68, 69
   communication
      services, 65–67
   digital advertising, 67, 68
   finance and investment
      industries, 62, 63
   GPT, 59
   healthcare and
      biotechnology, 60, 61
   healthcare crisis, 57–59
   information
      technology, 63–65
   legal document creation, 67
productivity, 403
prompt, 404
rights, 115
skills, 123, 124
transparency, 124–126
Generative AI hub, 394, 396, 397
Generative AI technologies, 9
Generative engine optimization
   (GEO), 22
GEO, *see* Generative engine
   optimization (GEO)
GitHub, 94
Google BARD, 23
Google search engine results page
   (SERP), 78
GPT-4 algorithm, 116

H

Health Insurance Portability and
   Accountability Act
   (HIPAA), 35
HIPAA, *see* Health Insurance
   Portability and
   Accountability Act (HIPAA)

I

IAM, *see* Identity and Access
   Management (IAM)
IDE, *see* Integrated development
   environment (IDE)

Identity and Access Management (IAM), 37
Improved susceptible-infected (ISI) model, 91
Integrated development environment (IDE), 217
Intellectual property (IP), 115
Intellectual property (IP) rights, 116
Intelligent agents (IAs), 8
Internet of things (IoT), 87
IP, *see* Intellectual property (IP)

J, K

Joint Research Agreement (JRA), 395
JRA, *see* Joint Research Agreement (JRA)

L

Lambda and API gateway
  Amazon S3, 326
  author from scratch, select, 340
  bedrock.py file, 352–354
  bucket, name, 328–330
  click bucket, 331
  configuration, 359
  create bucket button, 327
  create button, 338
  create folder, 331–333
  create function button, 341
  create functions, 339
  create layer, 336
  deploy button, 341–345
  download Postman, 371, 373
  execute scripts, 321
  functions, 320
  __init__.py file, 350, 351
  JSON, 362, 363, 365, 367
  layer configuration, 336
  layers, 356–358
  myutils folder, 349
  download myutils.zip, 324
  permissions, 361, 362
  rename button, 324
  rename notebook, 323
  REST API, 369, 370
  select button, 322
  terminal button, 334, 335
  textToSQL function, 348
  timeout, 360
Langchain SQL agent, 268–272, 274, 275, 277, 278, 280, 282, 283
Language modeling, 14
Large language models (LLMs), 55, 86, 173, 395, 406
LIME, *see* Local Interpretable Model-agnostic Explanations (LIME)
LLMs, *see* Large language models (LLMs)
Local Interpretable Model-agnostic Explanations (LIME), 124

INDEX

Long short-term memory
(LSTM), 90
LSTM, see Long short-term
memory (LSTM)

M

Machine learning (ML), 2
algorithms, 86
models, 248
Massachusetts Institute of
Technology (MIT), 3
Microsoft, 78
Microsoft Copilot, 80
Midjourney, 26, 94
Minimal viable products
(MVPs), 96
MIT, see Massachusetts
Institute of
Technology (MIT)
ML, see Machine learning (ML)
MLP, see Multilayer
perceptron (MLP)
Modern industries, GenAI
chest CI images, 87, 88
chest X-rays, 88
COVID-19 pandemic, 88–90
cyborgs, 97–102
drug discovery, 91
drug repurposing, 91
hybrid world, 94, 95
vaccine development, 92
vaccine trials, 93
Multilayer perceptron (MLP), 91

MVPs, see Minimal viable
products (MVPs)

N

Natural language processing
(NLP), 2–4, 107, 113, 405
NLP, see Natural language
processing (NLP)

O

OpenAI, 27
OpenAI's GPT, 25

P, Q

PandasAI, 373
POCs, see Proof of concepts (POCs)
Prompt engineering
advantages, 104, 105
definition, 102
LLMs, 103
practices, 111, 112
prompt, 103
techniques, 107–110
use cases, 105, 106
Proof of concepts (POCs), 96

R

RAG, see Retrieval augmented
generation (RAG)
Retrieval augmented generation
(RAG), 396

S

SAP BTP, *see* SAP Business Technology Platform (SAP BTP)
SAP business AI
    ERP/finance, 391
    human resources, 391
    industries, 392
    IT and platforms, 392, 393
    joule, 390
    quickly sorting and contextualizing data, 389
    sourcing/procurement, 394
    supply chain, 394
SAP Business Technology Platform (SAP BTP), 48
SAP Datasphere system, 401
SAP HANA Cloud, 286
    choose file, 249
    Claude LLM, 249
    machine learning models, 248
    project, 249–257, 259, 261, 262, 264–267
    query, 236, 237, 239–244, 246–248
    troubleshoot errors, 249
SAP HANA/PandasAI
    choose file, 301
    Cloud instance, 312
    code, 306–308
    connection established, 313
    data analysis workflows, 300
    drag and drop the file, 305
    download myutils.zip, 304
    play button, 314, 315, 317–319
    rename button, 304
    rename notebook, 303
    SAP BTP cloud instance, 310
    SAP BTP Cockpit, 309
    select button, 302
SAP Interactive Value Journey, 398, 399
SAP reports
    Claude LLM, 287
    click play, 293–297
    execution, 298, 299
    myutils.zip, download, 290
    rename button, 290
    rename notebook, 289
    RFSSLD00.zip, 291
    root folder, 292, 293
SAP Road Map Explorer, 400
SEIR model, *see* Susceptible-Exposed-Infectious-Recovered (SEIR) model
SHAP, *see* SHAPley Additive Explanations (SHAP)
SHAPley Additive Explanations (SHAP), 124
Stable Diffusion FM model, 146
Stable Diffusion model, 149, 152, 155
Support vector regression (SVR), 91

Susceptible-Exposed-Infectious-Recovered (SEIR) model, 90
SVR, *see* Support vector regression (SVR)

T, U, V, W, X, Y, Z
Tabnine, 94
textToSQL function, 348

3D protein model, 92
Titan Image Generator G1 foundation model, 172
Titan Image Generator G1 model, 164
Titan Text G1–Express model, 142
Tree-of-thought prompting, 108